T0225759

SpringerBriefs in Physics

Series Editors

Balasubramanian Ananthanarayan, Centre for High Energy Physics, Indian Institute of Science, Bangalore, India

Egor Babaev, Physics Department, University of Massachusetts Amherst, Amherst, MA, USA

Malcolm Bremer, H H Wills Physics Laboratory, University of Bristol, Bristol, UK

Xavier Calmet, Department of Physics and Astronomy, University of Sussex, Brighton, UK

Francesca Di Lodovico, Department of Physics, Queen Mary University of London, London, UK

Pablo D. Esquinazi, Institute for Experimental Physics II, University of Leipzig, Leipzig, Germany

Maarten Hoogerland, University of Auckland, Auckland, New Zealand

Eric Le Ru, School of Chemical and Physical Sciences, Victoria University of Wellington, Kelburn, Wellington, New Zealand

Dario Narducci, University of Milano-Bicocca, Milan, Italy

James Overduin, Towson University, Towson, MD, USA

Vesselin Petkov, Montreal, QC, Canada

Stefan Theisen, Max-Planck-Institut für Gravitationsphysik, Golm, Germany

Charles H.-T. Wang, Department of Physics, The University of Aberdeen, Aberdeen, UK

James D. Wells, Physics Department, University of Michigan, Ann Arbor, MI, USA

Andrew Whitaker, Department of Physics and Astronomy, Queen's University Belfast, Belfast, UK

SpringerBriefs in Physics are a series of slim high-quality publications encompassing the entire spectrum of physics. Manuscripts for SpringerBriefs in Physics will be evaluated by Springer and by members of the Editorial Board. Proposals and other communication should be sent to your Publishing Editors at Springer.

Featuring compact volumes of 50 to 125 pages (approximately 20,000–45,000 words), Briefs are shorter than a conventional book but longer than a journal article. Thus, Briefs serve as timely, concise tools for students, researchers, and professionals. Typical texts for publication might include:

- A snapshot review of the current state of a hot or emerging field
- A concise introduction to core concepts that students must understand in order to make independent contributions
- An extended research report giving more details and discussion than is possible in a conventional journal article
- A manual describing underlying principles and best practices for an experimental technique
- An essay exploring new ideas within physics, related philosophical issues, or broader topics such as science and society

Briefs allow authors to present their ideas and readers to absorb them with minimal time investment. Briefs will be published as part of Springer's eBook collection, with millions of users worldwide. In addition, they will be available, just like other books, for individual print and electronic purchase. Briefs are characterized by fast, global electronic dissemination, straightforward publishing agreements, easy-to-use manuscript preparation and formatting guidelines, and expedited production schedules. We aim for publication 8–12 weeks after acceptance.

More information about this series at http://www.springer.com/series/8902

M. S. Harazdyuk · V. T. Bachinsky ·
O. Ya. Wanchulyak · A. G. Ushenko ·
Yu. A. Ushenko · A. V. Dubolazov · M. P. Gorsky ·
A. V. Bykov · I. Meglinski

Correlation and Autofluorescence Microscopy in Forensics Medicine: Time of Death Detection Using Polycrystalline Cerebrospinal Fluid Films

 Springer

M. S. Harazdyuk
Department of Forensic Medicine
and Medical Law
Bukovinian State Medical University
Chernivtsi, Ukraine

O. Ya. Wanchulyak
Department of Forensic Medicine
and Medical Law
Bukovinian State Medical University
Chernivtsi, Ukraine

Yu. A. Ushenko
Yuriy Fedkovych Chernivtsi National
University
Chernivtsi, Ukraine

M. P. Gorsky
Yuriy Fedkovych Chernivtsi National
University
Chernivtsi, Ukraine

I. Meglinski
College of Engineering and Physical
Sciences
Aston University, Birmingham, UK

V. T. Bachinsky
Department of Forensic Medicine
and Medical Law
Bukovinian State Medical University
Chernivtsi, Ukraine

A. G. Ushenko
Yuriy Fedkovych Chernivtsi National
University
Chernivtsi, Ukraine

A. V. Dubolazov
Yuriy Fedkovych Chernivtsi National
University
Chernivtsi, Ukraine

A. V. Bykov
Opto-Electronics and Measurement
Techniques
University of Oulu, Oulu, Finland

Acknowledgement: National Research Foundation of Ukraine, Project 2020.02/0061.

ISSN 2191-5423   ISSN 2191-5431   (electronic)
SpringerBriefs in Physics
ISBN 978-981-16-0196-5   ISBN 978-981-16-0197-2   (eBook)
https://doi.org/10.1007/978-981-16-0197-2

© The Author(s), under exclusive licence to Springer Nature Singapore Pte Ltd. 2021
This work is subject to copyright. All rights are solely and exclusively licensed by the Publisher, whether the whole or part of the material is concerned, specifically the rights of translation, reprinting, reuse of illustrations, recitation, broadcasting, reproduction on microfilms or in any other physical way, and transmission or information storage and retrieval, electronic adaptation, computer software, or by similar or dissimilar methodology now known or hereafter developed.
The use of general descriptive names, registered names, trademarks, service marks, etc. in this publication does not imply, even in the absence of a specific statement, that such names are exempt from the relevant protective laws and regulations and therefore free for general use.
The publisher, the authors and the editors are safe to assume that the advice and information in this book are believed to be true and accurate at the date of publication. Neither the publisher nor the authors or the editors give a warranty, expressed or implied, with respect to the material contained herein or for any errors or omissions that may have been made. The publisher remains neutral with regard to jurisdictional claims in published maps and institutional affiliations.

This Springer imprint is published by the registered company Springer Nature Singapore Pte Ltd.
The registered company address is: 152 Beach Road, #21-01/04 Gateway East, Singapore 189721, Singapore

Contents

Abbreviations and Conventions

AOD   Antiquity of the onset of death
BBB   Blood–brain barrier
BF   Biological fluid
BT   Biological tissue
CCD   Charge-coupled device
CDMP   Complex degree of mutual polarization
CHD   Coronary heart disease
CSF   Cerebrospinal fluid
CT   Computer tomography (CT scan)
FTIS   Fourier-transform infrared spectroscopy
GABA   Gamma aminobutyric acid
LP   Laser polarimetry
NAD   Nicotinamide adenine dinucleotide
NADR   Nicotinamide adenine dinucleotide (reduced)
PFCSF   Polycrystalline films of cerebrospinal fluid
TSH   Thyroid stimulating hormone
VB   Vitreous body (vitreous)

Chapter 1
Materials and Methods

1.1 Substantiation of the Model of the Object of Study

The objects of the study are CSF samples taken from deceased and healthy volunteers. Such biological objects are complex optically heterogeneous structures [1–7].

Polycrystalline films are characterized by the simultaneous presence of optically isotropic and anisotropic components. In accordance with this approach, a wide range of mechanisms for converting the parameters of laser radiation passing through such a biological layer is realized. The mechanisms of the first group include optically isotropic reflection, refraction, and absorption. The mechanisms of the second group include local optically anisotropic interactions of two types:

1. optical activity—rotation of the plane of polarization of laser radiation by chiral molecules (albumin, globulin, glucose molecules);
2. linear birefringence—the formation of spatially ordered networks of protein molecules or needle crystals of albumin, fibrin, or elliptically polarized laser waves.

The result is a complex polarization-inhomogeneous optical image of PFCSF.

Separately, a third group of mechanisms is the phenomenon of fluorescence of molecular compounds (proteins, nicotinamide-adenine-dinucleotide (NAD), flavin, porphyrins), which are part of PFCSF, under the influence of shortwave laser radiation.

In our work, we studied the manifestations of all three groups of mechanisms during the passage of laser radiation through PFCSF, to identify patterns of postmortem changes in the polarization, polarization-correlation, and fluorescence structures of microscopic images [8–12].

© The Author(s), under exclusive licence to Springer Nature Singapore Pte Ltd. 2021   1
M. S. Harazdyuk et al., *Correlation and Autofluorescence Microscopy
in Forensics Medicine: Time of Death Detection Using Polycrystalline
Cerebrospinal Fluid Films*, SpringerBriefs in Physics,
https://doi.org/10.1007/978-981-16-0197-2_1

Table 1.1 Groups of deceased depending on the AOD and the number of samples studied

Group	2	3	4	5	6	7	8	9
AOD, год	2 ± 0.5	3 ± 0.6	6 ± 1.0	8 ± 1.0	12 ± 1.0	16 ± 1.0	18 ± 1.0	26 ± 2.0
Number of samples CSF	4	5	5	5	4	4	4	3

1.2 Characterization of Research Objects

For the study, corpses of the dead from cardiovascular pathology with a known time of death were selected. The exclusion criteria were as follows: the presence of traumatic brain injury, diseases of the central nervous system, suspected stroke, the presence of systemic and chronic diseases, poisoning with an unknown substance, narcotic or potent drugs. All of the samples obtained were divided into groups: group 1—samples obtained from healthy individuals, and groups 2–9—deceased with different AOD. The AOD in hours and the number of samples at each time interval are given in Table 1.1.

Experimental cytological studies of the temporal dynamics of postmortem changes in the morphology of CSF cells were performed according to the following algorithm:

1. CSF was taken by suboccipital puncture from a large occipital cistern in corpses and during spinal anesthesia in preparation for surgery in living healthy volunteers.
2. CSF films were formed under identical conditions by applying a drop to optically homogeneous glass immediately after liquid intake and after centrifugation at a speed of 3000 rpm for 5 min, followed by drying of the formed film at room temperature ($t = 22$ °C).

1.3 Schemes of Experimental Studies of Microscopic Polarization and Autofluorescence Images of Experimental Samples of Cerebrospinal Fluid

Three basic microscopic schemes of experimental studies were used:

1. Two-dimensional polarimetry of PFCSF:

    - Stokes-correlometry—determination of the coordinate distributions of the values of the complex degree of mutual polarization, characterizing the consistency of the polarization states at different points of the microscopic image;

2. Fluorescence Stokes-polarimetry of PFCSF microscopic images.

1.3.1 Optical Scheme of Spectral-Selective Stokes-Polarimetry and Its Characteristics

The measurements of coordinate distributions (two-dimensional arrays of values in the plane of samples of polycrystalline CSF films) of the azimuth and polarization ellipticity at the points of microscopic images were performed at the location (Fig. 1.1) of a standard Stokes-polarimeter.

A detailed description and characteristics of this system are given in advance.

In our work, we will dwell on a brief description that is necessary for a better understanding of the material of experimental measurements [13–20].

The irradiation mode of polycrystalline films CSF, 6, consisted of two channels:

- polarimetry—a "red" beam parallel to a diameter of 2 mm with a wavelength of 0.63 μm of a semiconductor laser, 1;
- fluorometry—parallel to the diameter of a 2-mm beam of "blue" with a wavelength of 0.405 μm of a semiconductor laser, 1.

The parallel laser beam was formed using a collimator, 2. A quarter-wave plate, 3, and a polarizer, 4, formed a predetermined polarization state of the probe beam. Images of samples of polycrystalline CSF, 6, films using a polarizing microlens, 7, were projected into the photosensitive plane (number of pixels 1280 × 960) of a digital CCD camera, 10.

The optical analysis of microscopic images of samples of polycrystalline films of CSF, 6, was conducted using a quarter-wave plate, 8, and a polarizer, 9.

The main information on the objectives of the given scheme is as follows, in total:

- to directly measure the coordinate distributions of the azimuth (α) and ellipticity (β) of the polarization at the points of the microscopic images—polarization maps;
- to calculate the distributions of the polarization-correlation parameter—the complex degree of mutual polarization (CDMP), which characterizes the degree of correlation consistency of the polarization states of neighboring points of the PFCSF microscopic image—"polarization-correlation maps".

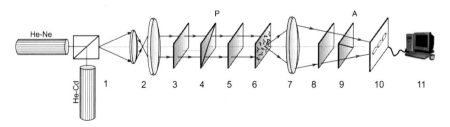

Fig. 1.1 Optical scheme of a Stokes-polarimeter with "two-wave" spectral-selective sounding channels. Explanation is in the text

1.3.2 Optical Scheme of Stokes-Polarimetry with Spatial-Frequency Filtering of Microscopic Images and the Characteristics of This Approach

Figure 1.2 presents an optical scheme for a laser Stokes-polarimeter with spatial-frequency filtering.

The main differences between this scheme and the classical Stokes-polarimeter are that the samples of polycrystalline films, 6, were placed in the focal plane of the polarizing microlens, 7. In the rear focal plane of the microlens, 7, was placed a spatial-frequency (low-frequency or high-frequency) filter-diaphragm, 8. A polarizing microlens, 9, was mounted at the focal length from the frequency plane of the lens, 7, and it realized the inverse Fourier transform of a spatial-frequency filtered laser field with a wavelength of 0.63 μm. The intensity of such a field was recorded by a digital CCD camera, 12, which was at the focal distance from the polarizing micro lens, 9.

Optical analysis of the microscopic images of samples of polycrystalline films of CSF, 6, was conducted using a quarter-wave plate, 10, and a polarizer, 11.

The main information on the objectives for the given scheme is as follows, in total:

- to directly measure the spatial-frequency filtered maps of the azimuth $F(\alpha)$ and ellipticity $F(\beta)$ of the polarization and CDMP.

Depending on the size of the aperture, 8, we will distinguish the following:

- "high-frequency maps", which characterize the small-scale structural dimensions of the microscopic images, in the range of 2–15 μm;
- "low-frequency maps", which characterize the large-scale structural sizes of the microscopic images, in the range of 20–100 μm.

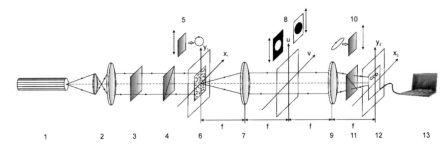

Fig. 1.2 Optical design of a Stokes-polarimeter using spatial-frequency filtering, where 1 is a He–Ne laser; 2 is a collimator; 3 is a stationary quarter-wave plate; 5, 10 are mechanically movable quarter-wave plates; 4, 11 are a polarizer and analyzer, respectively; 6 is polycrystalline film CSF; 7, 9 are polarizing microlenses; 8 is composed of low-pass and high-pass diaphragm-filters; 12 is a CCD camera; and 13 is a personal computer

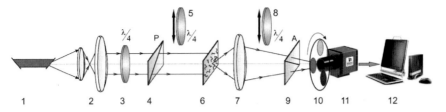

Fig. 1.3 Optical scheme of an autofluorescent Stokes-polarimeter. Explanation is in the text

1.3.3 Autofluorescence Laser Polarimetry

Figure 1.3 shows a diagram of a modified laser Stokes-polarimeter for autofluorescence studies of biological layers.

To excite autofluorescence in polycrystalline CSF films (Fig. 1.3), we used a "blue" semiconductor laser, LSR405ML-LSR-PS-II, with a wavelength of 0.405 μm and a power of 50 mW in the location of a standard Stokes-polarimeter.

The main information on the objectives of the given scheme is as follows, in total:

- to directly measure the autofluorescence intensity distributions I^{Φ} of polycrystalline CSF films;
- to obtain the azimuths of the polarization α^{Φ} of laser-induced fluorescence in the polycrystalline CSF layer.

1.3.4 Measurement of Polarization Maps of the Microscopic Image of PFCSF

This technique is described in detail in a series of research publications. We give a brief description, with specific data on our samples—polycrystalline CSF films. In the first stage, the coordinate distribution of the four parameters of the Stokes vector of the microscopic image is determined. For this purpose, six microscopic images are recorded using a polarizing filter, 8 and 9 (Fig. 1.4).

- We fix the transmission plane of the polarizer–analyzer, 9 (Fig. 1.4), at an angle of 0° relative to the plane of incidence, and we register the corresponding microscopic image of the PFCSF.
- We fix the plane of transmission of the polarizer 9 at an angle of 90°, and we register a microscopic image of PFCSF (Fig. 1.5).

Based on the determination of the Stocks vector S, we determine its first S_1 and second S_2 parameters by adding ($S_1 = I_0 + I_{90}$) and subtracting ($S_2 = I_0 - I_{90}$), as the intensity value in each pixel of the digital camera, 10 (Fig. 1.3).

Figures 1.6 and 1.7 show the coordinate distributions of the values of the 1st and 2nd parameters of the Stokes vector (upper parts) and the histograms of the random

Fig. 1.4
Polarized-microscopic image
(0°) of PFCSF

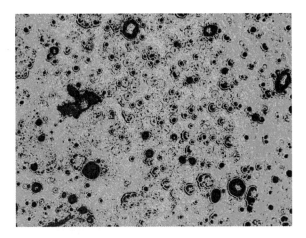

Fig. 1.5
Polarized-microscopic image
(90°) of PFCSF

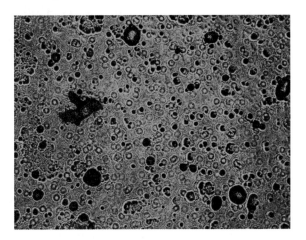

values of these parameters (lower parts) within the set of pixels of a digital camera
that records a microscopic image of the PFCSF.

- We fix the transmission plane of the polarizer, 9, at an angle of 45°, and we register
  the corresponding microscopic image of PFCSF.
- We fix the transmission plane of the polarizer, 9, at an angle of 135°, and we
  register the corresponding microscopic image of the PFCSF (Figs. 1.8 and 1.9).

By subtracting the intensity values ($S_3 = I_{45} - I_{135}$) in the pixels of the partial
microscopic images, we calculate the distribution of the values of the third parameter,
S_3, of the Stokes vector—Fig. 1.10.

- The distribution of the magnitude of the fourth parameter of the Stokes vector
  (S_4) was measured by sequential rotation of the axis of the quarter-wave plate, 8
  (Fig. 1.3), at the angles of 0° and 90° and the registration of the corresponding

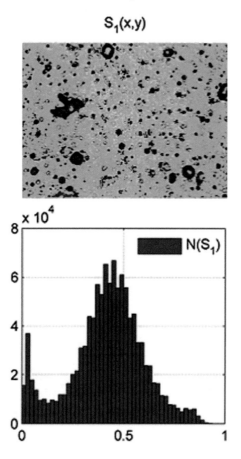

Fig. 1.6 Two-dimensional distribution and histogram of the values of the 1st parameter of the Stokes vector (S_1) of the microscopic image of PFCSF

microscopic images $\otimes$ and $\oplus$ with a subsequent subtraction of the intensities $S_4 = I_\otimes - I_\oplus$ (Figs. 1.11 and 1.12).

1.3.5 Methods of Experimental Measurement of the Degree of Mutual Polarization of Microscopic Images of Polycrystalline Films of Cerebrospinal Fluid

The technique of experimental measurements of coordinate distributions of a complex degree of mutual polarization (CDMP) has been presented in detail in a series of scientific papers. Here, we give a brief description.

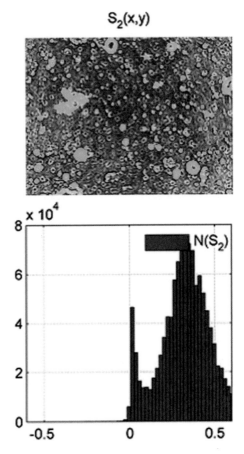

Fig. 1.7 Two-dimensional distribution and histogram of the values of the 2nd parameter of the Stokes vector (S_2) of the microscopic image of PFCSF

Fig. 1.8
Polarized-microscopic image
(45°) of PFCSF

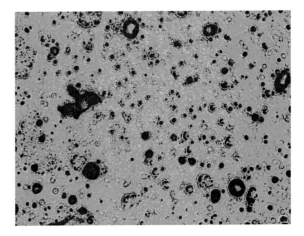

Fig. 1.9
Polarized-microscopic image
(135°) of PFCSF

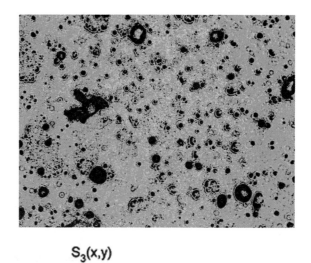

$$S_3(x,y)$$

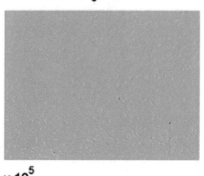

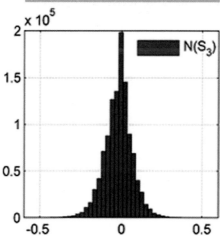

Fig. 1.10 Two-dimensional distribution and histogram of the values of the 3rd parameter of the
Stokes vector (S_3) of the microscopic image of PFCSF

Fig. 1.11
Polarized-microscopic image
of PFCSF

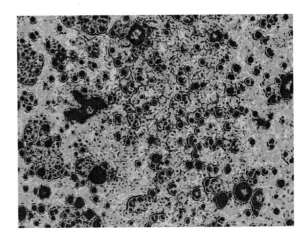

Fig. 1.12
Polarized-microscopic image
of PFCSF

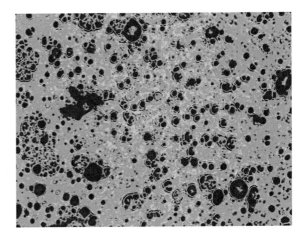

Firstly, a digital camera, 10, measures (in the absence of a polarizer-analyzer, 8, Fig. 2.1) the coordinate distribution of the intensity value at the points of a digital microscopic image of PFCSF—a classic microscopic image.

Secondly, by rotating the transmission plane of the polarizer, 8, and the axis of the maximum speed of the quarter-wave plate, 9 (Fig. 1.3), we measure the coordinate maps of the Stokes vector parameters (Figs. 1.6, 1.7, 1.8 and 1.13) and calculate the maps of the azimuth and ellipticity of the polarization of the digital microscopic image of PFCSF.

Thirdly, using the formula for determining the magnitude of the phase shift between the orthogonal components of the amplitude of the laser radiation, the phase map of the microscopic image of the PFCSF is calculated.

Fourth, the coordinate distribution of the CDMP magnitude of the microscopic image of PFCSF is determined by an algorithmic combination of the intensity,

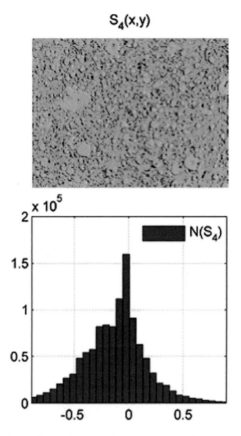

Fig. 1.13 Two-dimensional distribution and histogram of the values of the 4th parameter of the Stokes vector S_4 of the microscopic image of PFCSF

azimuth and polarization ellipticity maps and phase shifts. The specific form of this algorithm has been given previously.

The series Figs. 1.14, 1.15 and 1.16 shows examples of direct and spatial-frequency determination of the coordinate structure of the CDMP and a histogram of the distribution of their values.

Tables 1.2, 1.3 and 1.4 show the values of the statistical moments of the 1st–4th orders, which characterize the distribution of random values of the CDMP points of the microscopic images of polycrystalline films of CSF.

1.3.6   Polarization-Correlation Map

An analysis of the data showed that the statistical moments of the 3rd (2.32-fold increase) and 4th (1.53-fold increase) orders, which characterize the skewness and

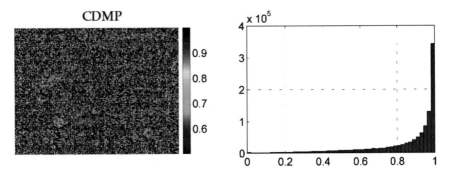

Fig. 1.14 Two-dimensional CDMP map of the microscopic image of PFCSF (left side) and a histogram of the distribution of its random values (right side)

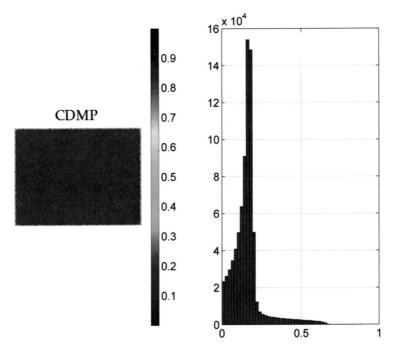

Fig. 1.15 Two-dimensional CDMP map of the small-scale component of the microscopic image of PFCSF (left side) and a histogram of the distribution of its random values (right side)

kurtosis of the histograms, are the largest compared with the results of mapping the polarization maps of the azimuth and ellipticity for the range of changes in the distribution of random values of the CDMP microscopic image of PFCSF.

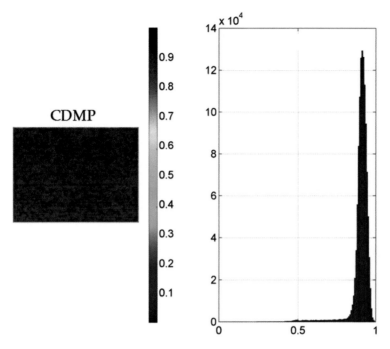

Fig. 1.16 Two-dimensional CDMP map of the large-scale component of the microscopic image of PFCSF (left side) and a histogram of the distribution of its random values (right side)

Table 1.2 Statistical moments' 1st–4th orders, which characterize the distribution of the CDMP in the microscopic images of PFCSF

Z_i	15 min	6 h
Z_1	0.92	0.65
Z_2	0.09	0.06
Z_3	0.52	1.21
Z_4	1.33	2.04

Table 1.3 Statistical moments of the 1st–4th orders, which characterize the CDMP distributions of the small-scale component of the microscopic images of PFCSF

Z_i	15 min	6 h
Z_1	0.19	0.13
Z_2	0.11	0.08
Z_3	0.79	1.68
Z_4	0.64	1.52

Table 1.4 Statistical moments of the 1st–4th orders, which characterize the CDMP distributions of the large-scale components of the microscopic images of PFCSF

Z_i	15 min	6 h
Z_1	0.91	0.93
Z_2	0.08	0.06
Z_3	0.31	0.59
Z_4	0.43	0.54

1.3.7 High-Frequency Polarization-Correlation Map

The use of spatial-frequency filtering of polarization-inhomogeneous microscopic images has increased the sensitivity of polarization-correlation mapping of small-scale optically anisotropic molecular compounds in PFCSF. Quantitatively, this finding illustrates growth in the skewness (growth of 2.12 times) and kurtosis (growth of 2.38 times), which characterize the distribution of the CDMP values.

1.3.8 Low-Frequency Polarization-Correlation Map

Statistical analysis of the coordinate distributions of the CDMP values of the microscopic image of large-scale needle-shaped crystalline formations of the CSF film revealed a significant sensitivity of temporal monitoring compared with the methods of azimuth and ellipticity polarimetry: the third-order statistical moment increases 1.9 times, and the fourth-order statistical moment increases 1.26 times.

Thus, mapping the distributions of the CDMP values was determined to be the most promising approach for the temporary monitoring of classical microscopic images, as well as images of the small-scale structure of optically active polycrystalline networks of CSF films.

1.3.9 Method of Two-Dimensional Mapping of Autofluorescence of Polycrystalline Films of Cerebrospinal Fluid

The coordinate distributions of the values of the intensity of laser-induced autofluorescence of molecular compounds of polycrystalline SMR films, 6, in the plane of the photosensitive area of the digital camera, 11, were measured using a set of bandpass interference filters, 10 (Fig. 1.3), with spectral transmission maxima in the following spectral ranges:

- "blue"—transmission center at a wavelength of 0.45 μm, where the maximum fluorescence of proteins and nicotinamide nucleotide (NADR) are localized;
- "yellow-green"—transmission center at a wavelength of 0.55 microns, where the fluorescence maxima of flavins, folic acids are localized;
- "red"—transmission center at a wavelength of 0.63 microns, where the fluorescence maxima of porphyrin are localized.

The distribution of the azimuth of the polarization azimuth of the laser-induced fluorescence PFCSF was measured by the method described previously.

The results of two-dimensional mapping in different spectral ranges of the distribution of the intensities and azimuths of the polarization of autofluorescence of polycrystalline CSF films are illustrated by the series in Figs. 1.17, 1.18, 1.19, 1.20, 1.21 and 1.22.

Tables 1.5, 1.6, 1.7 and 1.8 show the statistical parameters that characterize the distributions of the values of the intensity and azimuth of the polarization in microscopic images of autofluorescence in polycrystalline CSF films.

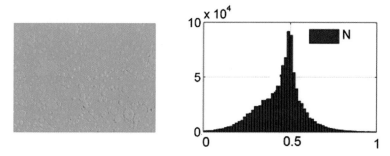

Fig. 1.17 Map of the fluorescence intensity in the "blue" region of the PFCSF spectrum (left side) and a histogram of the distribution of its random values (right side)

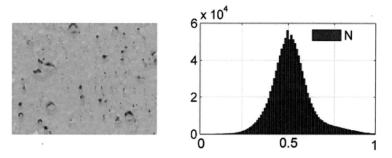

Fig. 1.18 Map of the fluorescence intensity of PFCSF (left side) and histogram of the distribution of its values (right side) in the yellow-green region of the spectrum

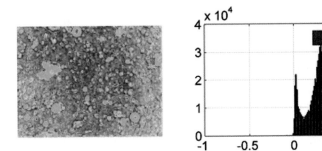

Fig. 1.19 Map of the fluorescence intensity of PFCSF (left side) and histogram of the distribution of its values (right side) in the red region of the spectrum

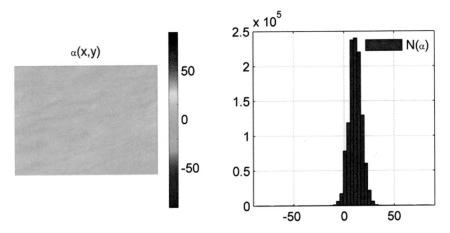

Fig. 1.20 Two-dimensional map of the azimuth of the polarization of fluorescence of PFCSF (left side) and a histogram of the distribution of its random values (right side) in the blue region of the spectrum

1.3.10 Fluorescence of the Proteins, Nicotinamide-Dinucleotides (NADR)

In the "blue" part of the spectrum of intrinsic laser-induced fluorescence of PFCSF, the following regularities of the temporal changes in the values of the statistical moments with 1–4 order of magnitude, which characterized the intensity distribution, were revealed:

- average—decreases by 1.51 times;
- dispersion—decreases by 1.44 times;
- skewness grows 2.18 times;
- kurtosis grows 2.53 times.

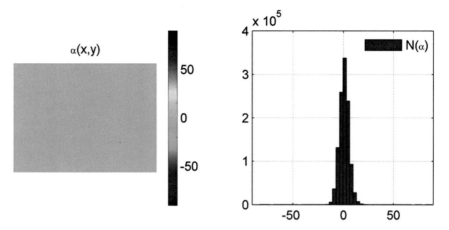

Fig. 1.21 Two-dimensional map of the azimuth of the polarization of the fluorescence of PFCSF (left side) and a histogram of the distribution of its random values (right side) in the yellow-green region of the spectrum

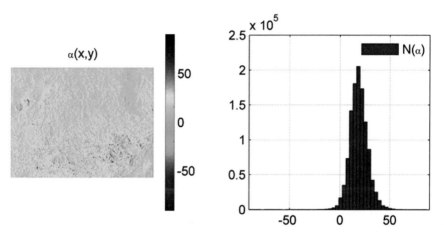

Fig. 1.22 Two-dimensional map of the azimuth of the polarization of the fluorescence of PFCSF (left side) and a histogram of the distribution of its random values (right side) in the red region of the spectrum

Table 1.5 Statistical moments of the 1st–4th orders, which characterize the distribution of the fluorescence intensity of the polycrystalline films of CSF of the donor (15 min) and the deceased (6 h) in the blue region of the spectrum

Z_i	15 min	6 h
Z_1	0.41	0.27
Z_2	0.13	0.09
Z_3	0.38	0.83
Z_4	0.43	1.09

Table 1.6 Statistical moments of the 1st–4th orders, which characterize the distribution of the fluorescence intensity of the polycrystalline films of CSF of the donor (15 min) and the deceased (6 h) in the yellow-green region of the spectrum

Z_i	15 min	6 h
Z_1	0.44	0.23
Z_2	0.15	0.12
Z_3	0.21	0.83
Z_4	0.19	0.43

Table 1.7 Statistical moments of 1st–4th orders, which characterize the distribution of the fluorescence intensity of polycrystalline films of the CSF of the donor (15 min) and the deceased (6 h) in the red region of the spectrum

Z_i	15 min	6 h
Z_1	0.39	0.13
Z_2	0.17	0.14
Z_3	0.78	1.06
Z_4	0.83	1.39

Table 1.8 Statistical moments of the 1st–4th orders, which characterize the distribution of the azimuths of the polarization of the fluorescence of polycrystalline films of CSF in the blue spectral region

Z_i	15 min	6 h
Z_1	0.11	0.07
Z_2	0.09	0.05
Z_3	0.63	1.11
Z_4	0.79	1.83

As seen, the statistical moments of the second and fourth orders are the most sensitive to the process of quenching the fluorescence of the protein molecules.

1.3.11 Fluorescence of Flavins, Folic Acids

In the "yellow-green" region of the spectrum of intrinsic laser-induced fluorescence PFCSF, the following regularities of the temporal changes in the values of the statistical moments of the 1st–4th orders, which characterize the intensity distribution, were revealed: average—decreases by 1.91 times; dispersion—decreases by 1.25 times; skewness—grows 3.95 times; kurtosis grows 2.26 times.

As seen, the range of the change and the sensitivity to the processes of quenching of fluorescence are commensurate with the data from temporary monitoring in the blue region of the spectrum.

1.3.12 Porphyrin Fluorescence

An analysis of the dynamics of the changes in the values of the statistical moments of the 1st–4th orders, which characterize the fluorescence images of the PFCSF at the different times of observation, revealed the highest sensitivity of the "red" spectral range:

- average—decreases by 3 times;
- dispersion—decreases by 1.21 times;
- skewness—grows by 1.35 times;
- kurtosis—grows by 1.67 times.

As seen, the statistical moments of the 1st, 3rd and 4th orders are the most sensitive to the process of quenching the fluorescence of porphyrias.

Additional possibilities in terms of improving the sensitivity and, accordingly, the accuracy of the temporary monitoring of biochemical changes in PFCSF were revealed by the method of laser-induced fluorescence polarimetry of the azimuth, as shown in Figs. 1.20, 1.21, 1.22 and Tables 1.8, 1.9, 1.10.

Table 1.9 Statistical moments of the 1st–4th orders, which characterize the distribution of the azimuth values of the polarization of the fluorescence of polycrystalline films of CSF in the yellow-green spectral region

Z_i	15 min	6 h
Z_1	0.09	0.05
Z_2	0.06	0.04
Z_3	0.48	0.89
Z_4	0.81	1.22

Table 1.10 Statistical moments of the 1st–4th orders, which characterize the distribution of the azimuth values of the polarization of the fluorescence of polycrystalline films of CSF in the red spectral region

Z_i	15 min	6 h
Z_1	0.14	0.09
Z_2	0.12	0.08
Z_3	0.35	0.81
Z_4	0.59	1.27

1.3.13   Polarization Maps of Fluorescence of Proteins and NADRs

In the short-wavelength "blue" region of the spectrum of the intrinsic laser-induced fluorescence of PFCSF, the following regularities of the temporal changes in the statistical moments of the 1st–4th orders, which characterize the distribution of the values of the azimuth of the polarization, are revealed as follows:

- average—decreases by 1.57 times;
- dispersion—decreases by 1.8 times;
- skewness grows 1.76 times;
- kurtosis grows 2.32 times.

A comparative analysis with data from a statistical analysis of the distribution of intrinsic fluorescence intensity in this spectral region revealed an increase in the sensitivity to temporary changes in the optical anisotropy of PFCSF. Moreover, all of the statistical moments of the 1st–4th orders were determined to be the most sensitive to the processes involved in quenching the fluorescence of protein molecules.

1.3.14   Polarization Maps of the Fluorescence of Flavins and Folic Acids

It has been established that in the "yellow-green" region of the spectrum of intrinsic laser-induced fluorescence of PFCSF, the following temporary changes in the values of the statistical moments of the 1st–4th orders, which characterize the distribution of the values of the azimuth of the polarization, take place: average—decreases by 1.8 times; dispersion—decreases by 1.5 times; skewness grows by 1.85 times; kurtosis grows 1.51 times. A comparative analysis of the statistical analysis of the distribution of the intensity of the intrinsic fluorescence in this region of the spectrum revealed the statistical moments that are the most sensitive to fluorescence quenching were the average and dispersion.

1.3.15   Polarization Maps of the Fluorescence of Porphyrins

In analysis of the temporal dynamics of the changes in the values of the statistical moments of the 1st–4th orders, which characterize the polarization maps of the azimuths of the microscopic images of laser-induced fluorescent PFCSFs at different times of observation, the sensitivity of the "red" spectral range revealed a similar sensitivity as in the short-wavelength "blue" region of the spectrum:

- average—decreases by 1.56 times;

- dispersion—decreases by 1.5 times;
- skewness grows 2.31 times;
- kurtosis grows 2.15 times.

As seen, the statistical moments of the 3rd and 4th orders are the most sensitive to the processes of quenching the fluorescence of porphyrins.

1.4 Algorithm for Determination of the Time of Death Onset

This section proposes a new algorithm for the analytical determination of the time of the onset of death.

The common denominator of the analysis and the justification of the above experimental data was determined to be the almost linear nature of the temporary change in the magnitude of the statistical moments of the 1st–4th orders, which characterize the distribution of objective parameters of postmortem laser microscopic images of histological sections of biological tissues. This fact is the basis for obtaining an analytical algorithm for determining and increasing the accuracy of AOD determination while maintaining the value of the time intervals based on monitoring changes in the statistical structure of postmortem microscopic images of polycrystalline films of CSF.

The analytical determination scheme of AOD is illustrated in Fig. 1.23.

From the analysis of the model scheme, the following relationships can be obtained to determine the AOD:

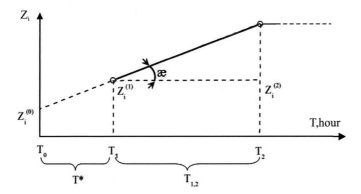

Fig. 1.23 Analysis of the algorithm for the analytical determination of AOD. Here: T_1—time of the start of the measurement of the indicator $Z_i^{(1)}$; T_2—the time of the completion of the measurement of the indicator $Z_i^{(2)}$ at the stage of "stabilization" of the change in its value ($Z_i^{(2)}(T) \approx const$); T_0—death time; ξ—angle of information dependence $Z_i(T)$

Table 1.11 Statistical moments of the 1st–4th orders, which characterize the polarization-correlation maps of the CDMP obtained in the lifetime of the polycrystalline films of CSF

Z_i^0	W	W^*	W^{**}
Z_1^0	0.92 ± 0.078	0.19 ± 0.014	0.91 ± 0.077
p	$p \prec 0.001$	$p \prec 0.001$	$p \prec 0.001$
Z_2^0	0.09 ± 0.008	0.11 ± 0.009	0.08 ± 0.006
p	$p \prec 0.001$	$p \prec 0.001$	$p \prec 0.001$
Z_3^0	0.93 ± 0.041	0.53 ± 0.064	1.02 ± 0.023
p	$p \prec 0.001$	$p \prec 0.001$	$p \prec 0.001$
Z_4^0	1.69 ± 0.11	0.35 ± 0.053	0.74 ± 0.12
p	$p \prec 0.001$	$p \prec 0.001$	$p \prec 0.001$

Notes: W—CDMP value in the plane of the classical digital microscopic images of polycrystalline films of CSF

W^*—CDMP value in the plane of the high-frequency (small-scale) component of the digital microscopic images of polycrystalline films of CSF

W^{**}—CDMP value in the plane of the low-frequency (large-scale) component of the digital microscopic images of polycrystalline films of CSF

$$tg\xi = \frac{Z_i^{(2)} - Z_i^{(1)}}{T_2 - T_1} = \frac{\Delta Z_i^{(1,2)}}{\Delta T_{12}}. \tag{1.1}$$

Using relation (1.1), we obtain an expression for determining the AOD

$$T^* = \left(Z_i^{(1)} - Z_i^{(0)}\right)\frac{\Delta T_{12}}{\Delta Z_i^{(1,2)}}. \tag{1.2}$$

Here, $Z_i^{(0)}$ is the value of the objective parameter determined by the intake sample of the CSF during the life of the donors (Tables 1.2, 1.3, 1.4, 1.5, 1.6 and 1.7).

Tables 1.11, 1.12 and 1.13 show the average values $\tilde{Z}_i^{(0)}$ and deviations $\pm\sigma$ within a statistically significant data sampling of samples obtained in the lifetime of the polycrystalline films of CSF.

Table 1.12 Statistical moments of 1st–4th orders, which characterize the fluorescence intensity maps obtained in the lifetime of CSF films

Z_i^0	$I_\Phi (\lambda = 0.45\,\mu m)$	$I_\Phi (\lambda = 0.55\,\mu m)$	$I_\Phi (\lambda = 0.63\,\mu m)$
Z_1^0	0.36 ± 0.034	0.43 ± 0.036	0.49 ± 0.031
p	$p \prec 0.001$	$p \prec 0.001$	$p \prec 0.001$
Z_2^0	0.13 ± 0.099	0.15 ± 0.011	0.17 ± 0.013
p	$p \prec 0.001$	$p \prec 0.001$	$p \prec 0.001$
Z_3^0	0.48 ± 0.023	0.46 ± 0.031	0.27 ± 0.016
p	$p \prec 0.001$	$p \prec 0.001$	$p \prec 0.001$
Z_4^0	0.78 ± 0.041	0.27 ± 0.014	0.74 ± 0.039
p	$p \prec 0.001$	$p \prec 0.001$	$p \prec 0.001$

Notes: $I_\Phi (\lambda = 0.45\,\mu m)$—the value of the fluorescence intensity in the "blue" ($\lambda = 0.45\,\mu m$) range of the spectrum

$I_\Phi (\lambda = 0.55\,\mu m)$—the value of the fluorescence intensity in the "green" ($\lambda = 0.55\,\mu m$) range of the spectrum

$I_\Phi (\lambda = 0.63\,\mu m)$—the value of the fluorescence intensity in the "red" ($\lambda = 0.63\,\mu m$) range of the spectrum

Table 1.13 Statistical moments of the 1st- to 4-th orders, which characterize the azimuth maps of the polarizations obtained in the lifetime of the CSF films

Z_i^0	$\alpha_\Phi (\lambda = 0.45\,\mu m)$	$\alpha_\Phi (\lambda = 0.55\,\mu m)$	$\alpha_\Phi (\lambda = 0.63\,\mu m)$
Z_1^0	0.11 ± 0.08	0.09 ± 0.007	0.14 ± 0.01
p	$p \prec 0.001$	$p \prec 0.001$	$p \prec 0.001$
Z_2^0	0.09 ± 0.007	0.06 ± 0.004	0.12 ± 0.009
p	$p \prec 0.001$	$p \prec 0.001$	$p \prec 0.001$
Z_3^0	0.63 ± 0.041	0.59 ± 0.032	0.52 ± 0.028
p	$p \prec 0.001$	$p \prec 0.001$	$p \prec 0.001$
Z_4^0	0.81 ± 0.057	0.94 ± 0.062	0.76 ± 0.041
p	$p \prec 0.001$	$p \prec 0.001$	$p \prec 0.001$

Notes: $\alpha_\Phi (\lambda = 0.45\,\mu m)$—the value of the azimuth of the fluorescence polarization in the "blue" ($\lambda = 0.45\,\mu m$) range of the spectrum

$\alpha_\Phi (\lambda = 0.55\,\mu m)$—the value of the azimuth of the fluorescence polarization in the "green" ($\lambda = 0.55\,\mu m$) range of the spectrum

$\alpha_\Phi (\lambda = 0.63\,\mu m)$—the value of the azimuth of the fluorescence polarization in the "red" ($\lambda = 0.63\,\mu m$) range of the spectrum

References

1. Ushenko, V.O., Vanchuliak, O., Sakhnovskiy, MYu., Dubolazov, O.V., Grygoryshyn, P., Soltys, I.V., Olar, O.V.: System of Mueller matrix polarization correlometry of biological polycrystalline layers. Proc. SPIE Int. Soc. Opt. Eng. **10352**, 103520U (2017)
2. Ushenko, V.O., Vanchuliak, O., Sakhnovskiy, M.Y., Dubolazov, O.V., Grygoryshyn, P., Soltys, I.V., Olar, O.V., Antoniv, A.: Polarization-interference mapping of biological fluids polycrystalline films in differentiation of weak changes of optical anisotropy. Proc. SPIE Int. Soc. Opt. Eng. **10396**, 103962O (2017)
3. Dubolazov, O.V., Trifonyuk, L., Marchuk, Y., Ushenko, Y.O., Zhytaryuk, V.G., Prydiy, O.G., Kushnerik, L., Meglinskiy, I.: Two-point Stokes vector parameters of object field for diagnosis and differentiation of optically anisotropic biological tissues. Proc. SPIE Int. Soc. Opt. Eng. **10352**, 103520V (2017)
4. Trifonyuk, L., Dubolazov, O.V., Ushenko, Y.O., Zhytaryuk, V.G., Prydiy, O.G., Grytsyuk, M., Kushnerik, L., Meglinskiy, I., Savka, I.G.: New opportunities of differential diagnosis of biological tissues polycrystalline structure using methods of Stokes correlometry mapping of polarization inhomogeneous images. Proc. SPIE Int. Soc. Opt. Eng. **10396**, 103962R (2017)
5. Dubolazov, O.V., Ushenko, V.O., Trifoniuk, L., Ushenko, Y.O., Zhytaryuk, V.G., Prydiy, O.G., Grytsyuk, M., Kushnerik, L., Meglinskiy, I.: Methods and means of 3D diffuse Mueller-matrix tomography of depolarizing optically anisotropic biological layers. Proc. SPIE Int. Soc. Opt. Eng. **10396**, 103962P (2017)
6. Ushenko, A.G., Dubolazov, A.V., Ushenko, V.A., Novakovskaya, O.Y.: Statistical analysis of polarization-inhomogeneous Fourier spectra of laser radiation scattered by human skin in the tasks of differentiation of benign and malignant formations. J. Biomed. Opt. **21**(7), 071110 (2016)
7. Ushenko, Y.A., Koval, G.D., Ushenko, A.G., Dubolazov, O.V., Ushenko, V.A., Novakovskaia, O.Y.: Mueller-matrix of laser-induced autofluorescence of polycrystalline films of dried peritoneal fluid in diagnostics of endometriosis. J. Biomed. Opt. **21**(7), 071116 (2016)
8. Prysyazhnyuk, V.P., Ushenko, YuA, Dubolazov, A.V., Ushenko, A.G., Ushenko, V.A.: Polarization-dependent laser autofluorescence of the polycrystalline networks of blood plasma films in the task of liver pathology differentiation. Appl. Opt. **55**(12), B126–B132 (2016)
9. Ushenko, A.G., Dubolazov, O.V., Ushenko, V.A., Novakovskaya, OYu., Olar, O.V.: Fourier polarimetry of human skin in the tasks of differentiation of benign and malignant formations. Appl. Opt. **55**(12), B56–B60 (2016)
10. Ushenko, Y.A., Bachynsky, V.T., Vanchulyak, O.Y., Dubolazov, A.V., Garazdyuk, M.S., Ushenko, V.A.: Jones-matrix mapping of complex degree of mutual anisotropy of birefringent protein networks during the differentiation of myocardium necrotic changes. Appl. Opt. **55**(12), B113–B119 (2016)
11. Dubolazov, A.V., Pashkovskaya, N.V., Ushenko, Y.A., Marchuk, Y.F., Ushenko, V.A., Novakovskaya, O.Y.: Birefringence images of polycrystalline films of human urine in early diagnostics of kidney pathology. Appl. Opt. **55**(12), B85–B90 (2016)
12. Garazdyuk, M.S., Bachinskyi, V.T., Vanchulyak, O.Ya., Ushenko, A.G., Dubolazov, O.V., Gorsky, M.P.: Polarization-phase images of liquor polycrystalline films in determining time of death. Appl. Opt. **55**(12), B67-B71 (2016)
13. Ushenko, A.G., Dubolazov, A.V., Ushenko, V.A., Ushenko, Y.A., Sakhnovskiy, M.Y., Olar, O.I.: Methods and means of laser polarimetry microscopy of optically anisotropic biological layers. Proc. SPIE Int. Soc. Opt. Eng. **9971**, 99712B (2016)
14. Ushenko, A.G., Dubolazov, A.V., Ushenko, V.A., Ushenko, Y.A., Kushnerick, L.Y., Olar, O.V., Pashkovskaya, N.V., Marchuk, Y.F.: Mueller-matrix differentiation of fibrillar networks of biological tissues with different phase and amplitude anisotropy Proc. SPIE Int. Soc. Opt. Eng. **9971**, 99712K (2016)
15. Dubolazov, O.V., Ushenko, A.G., Ushenko, Y.A., Sakhnovskiy, M.Y., Grygoryshyn, P.M., Pavlyukovich, N., Pavlyukovich, O.V., Bachynskiy, V.T., Pavlov, S.V., Mishalov, V.D.,

Omiotek, Z., Mamyrbaev, O.: Laser Mueller-matrix diagnostics of changes in the optical anisotropy of biological tissues. In: Information Technology in Medical Diagnostics II—Proceedings of the International Scientific Internet Conference on Computer Graphics and Image Processing and 48th International Scientific and Practical Conference on Application of Lasers in Medicine and Biology, vol. 2018, pp. 195–203 (2019)

16. Borovkova, M., Peyvasteh, M., Dubolazov, O., Ushenko, Y., Ushenko, V., Bykov, A., Deby, S., Rehbinder, J., Novikova, T., Meglinski, I.: Complementary analysis of Mueller-matrix images of optically anisotropic highly scattering biological tissues. J. Eur. Opt. Soc. **14**(1), 20 (2018)

17. Ushenko, V., Sdobnov, A., Syvokorovskaya, A., Dubolazov, A., Vanchulyak, O., Ushenko, A., Ushenko, Y., Gorsky, M., Sidor, M., Bykov, A., Meglinski, I.: 3D Mueller-matrix diffusive tomography of polycrystalline blood films for cancer diagnosis. Photonics **5**(4), 54 (2018)

18. Trifonyuk, L., Baranowski, W., Ushenko, V., Olar, O., Dubolazov, A., Ushenko, Y., Bodnar, B., Vanchulyak, O., Kushnerik, L., Sakhnovskiy, M.: 2D-Mueller-matrix tomography of optically anisotropic polycrystalline networks of biological tissues histological sections. Opto-Electron. Rev. **26**(3), 252–259 (2018)

19. Ushenko, V.A., Dubolazov, A.V., Pidkamin, L.Y., Sakchnovsky, M.Y., Bodnar, A.B., Ushenko, Y.A., Ushenko, A.G., Bykov, A., Meglinski, I.: Mapping of polycrystalline films of biological fluids utilizing the Jones-matrix formalism. Laser Phys. **28**(2), 025602 (2018)

20. Gavryliak, M.S., Prodan, D.I., Dubolazov, O.V., Gavryliak, D.S.: Spectral investigation of polarization properties of optical field scattered by muscle tissue. In: Proceedings of SPIE—The International Society for Optical Engineering, vol. 10750 (2018)

Chapter 2
Determination of the Time Onset of Death Based on a Statistical Analysis of the Distributions of the Values of the Complex Degrees of Mutual Polarization of Microscopic Images of Polycrystalline Films of Cerebrospinal Fluid

This chapter contains the results of studying the temporal dynamics of postmortem changes in the biochemical structure of polycrystalline CSF films by determining the magnitudes of the statistical moments of the 3rd and 4th orders, which characterize the coordinate distributions of the values of the complex degree of mutual polarization of various points:

- classic microscopic images of a polycrystalline network of biochemical crystals of CSF;
- a small-scale component of the microscopic image of an optically active molecular polycrystalline network of CSF biochemical crystals;
- a large-scale component of the microscopic image of a linearly birefringent polycrystalline network of CSF biochemical crystals.

For the set of coordinate distributions of the values of the complex degree of mutual polarization, the statistical moments of the 3rd and 4th orders were calculated. By summarizing the time dependences of the statistical moments of the 3rd and 4th orders, which characterize the postmortem changes in the polarization-correlation maps of the CDMP values of a series of microscopic images of polycrystalline networks of CSF films within statistically significant groups of samples, the intervals and the accuracy of determining the time of the onset of death are determined.

© The Author(s), under exclusive licence to Springer Nature Singapore Pte Ltd. 2021   27
M. S. Harazdyuk et al., *Correlation and Autofluorescence Microscopy in Forensics Medicine: Time of Death Detection Using Polycrystalline Cerebrospinal Fluid Films*, SpringerBriefs in Physics, https://doi.org/10.1007/978-981-16-0197-2_2

2.1 The Study of Temporal Post-mortem Changes in the Statistical Structure of the Complex Degree of Mutual Polarization of Microscopic Images of PFCSF

Previous diagnostic studies of the feasibility of establishing AOD using polarimetric mapping were based on the following regulations:

- each local point in the PFCSF plane has a certain level of optical anisotropy, which is manifested in a change in the polarization state of the laser beam irradiating the sample with subsequent rotation of the plane of polarization and the formation of its ellipticity;
- in accordance with the specific structure of the polycrystalline network of biochemical crystals, a polarization-inhomogeneous microscopic image is formed with coordinate-distributed values of the azimuth and polarization ellipticity at each point of such a CSF film image;
- polarization maps are "optical imprints" of the degree of crystallization of the polycrystalline network of biological crystals of CSF;
- polarimetric mapping of microscopic images allows the investigator to experimentally obtain information about the maps of azimuth and ellipticity in the form of a combination (distribution) of measurements taken at random;
- statistical analysis of two-dimensional distributions of azimuth and polarization ellipticity provides the ability to calculate a set of statistical moments of the 1st–4th orders for each time interval of postmortem PFCSF changes;
- the obtained time dependences of the postmortem changes in the magnitudes of the statistical moments are the basis for determining the time of the onset of death with a certain accuracy according to the analytical algorithm.

Summarizing the above, it can be stated that the basis for determining the AOD by all methods of polarimetric mapping is an array of random values of polarization parameters that do not have a clear coordinate position in the plane of the sample of the CSF film.

At the same time, by accounting for the coordinate location of various polarization states, one can obtain new, additional information on "subtle" postmortem changes in the biochemical composition of CSF. For this reason, the sensitivity and accuracy of the optical-physical determination of the time of the onset of death should increase.

2.2 Polarization-Correlation Mapping of Post-mortem Changes in the Complex Degree of Mutual Polarization of Microscopic Images of Optically Anisotropic Polycrystalline CSF Films

Experimental polarization-correlation studies of the temporal dynamics of the values of the complex degree of mutual polarization, which characterize the postmortem changes in the biochemical structure of CSF films, were conducted according to the following algorithm:

For each PFCSF sample in the optical location of the Stokes-polarimeter, the coordinate distributions of the values of the CDMP points in the plane of a polycrystalline film with a measurement step of 2 μm were determined by the method of measuring two-dimensional distributions of the values of the complex degree of mutual polarization. For each two-dimensional distribution of CDMP values of the set of points of the PFCSF microscopic image, we calculated the values of the statistical moments of the 1st–4th orders, which characterize the histograms of random CDMP values. Statistical processing of the measured set of values of the statistical moments of the 1st–4th orders, which characterize the distribution of CDMP values of the set of points of the PFCSF microscopic image, was conducted within a representative sample of samples until the confidence interval was reached at $p \prec 0.01$.

The time dependences of the changes in the values of the most sensitive statistical moments that characterize the distribution of CDMP values of the set of points of the PFCSF microscopic image were constructed, and the time was determined at which stabilization of the skewness and kurtosis of the CDMP histograms was achieved. Based on the information on the temporal dynamics of the changes in the CDMP distributions of the set of points of the PFCSF microscopic image, the interval T and accuracy ΔT of the AOD establishment were determined by applying the analytical method, a description of which is presented above (Tables 2.1, 2.2, 2.3 and 2.4).

Analysis of the data is revealed, as follows:

- Both polarization-correlation maps of the polycrystalline network of a CSF film, discretized within a set of pixels of digital microscopic images, indicate a significant coordinate heterogeneity of the optical manifestations of birefringence in different sections of the molecular crystals (Figs. 2.1 and 2.2, left parts) at different times of observation after death. Moreover, the rate of temporary postmortem changes in the structure of the network of biochemical crystals is significantly higher than in previous polarimetric studies of AOD.
- Quantitatively, the dynamics of the postmortem changes in the biochemical structure of the polycrystalline CSF films are characterized by both the range of changes in the values of the complex degree of mutual polarization in the histograms of the distribution of this parameter and the rate of decrease of the extreme values (Figs. 2.1 and 2.2, right-hand sides).

Table 2.1 Statistical moments of the 1st–4th orders, which characterize the distribution of the CDMP of the microscopic images of CSF polycrystalline films of all groups

Statistical moments	Control group (n = 30)	Cause of death (n = 30)	
	Donors	Myocardial infarction	Alcohol poisoning
Average, M_1	0.92 ± 0.056	0.61 ± 0.037	0.52 ± 0.029
P_1		$P < 0.05$	$P < 0.001$
P_2		$P > 0.05$	
Dispersion, M_2	0.09 ± 0.0054	0.06 ± 0.004	0.05 ± 0.003
P_1		$P < 0.05$	$P < 0.05$
P_2		$P > 0.05$	
Skewness, M_3	0.52 ± 0.031	1.22 ± 0.066	1.43 ± 0.081
P_1		$P < 0.001$	$P < 0.001$
P_2		$P > 0.05$	
Kurtosis, M_4	1.33 ± 0.081	2.23 ± 0.104	2.42 ± 0.13
P_1		$P < 0.001$	$P < 0.001$
P_2		$P > 0.05$	

Table 2.2 Statistical moments of the 1st–4th orders, which characterizing the distribution of the CDMP large-scale component of microscopic images of polycrystalline CSF films

Statistical moments	Control group	Cause of death	
	Donors (n = 30)	Myocardial infarction (n = 30)	Alcohol poisoning (n = 30)
Average, M_1	0.91 ± 0.056	0.59 ± 0.037	0.47 ± 0.029
P_1		$p < 0.05$	$p < 0.05$
P_2		$p > 0.05$	
Dispersion, M_2	0.08 ± 0.004	0.06 ± 0.003	0.05 ± 0.003
P_1		$p < 0.05$	$p < 0.05$
P_2		$p > 0.05$	
Skewness, M_3	0.31 ± 0.017	0.58 ± 0.034	0.67 ± 0.038
P_1		$p < 0.05$	$p < 0.05$
P_2		$p > 0.05$	
Kurtosis, M_4	0.43 ± 0.026	0.72 ± 0.034	0.84 ± 0.043
P_1		$p < 0.05$	$p < 0.05$
P_2		$p > 0.05$	

- With increasing observation time after death due to degenerative changes in polycrystalline CSF films, the optical anisotropy of protein molecules and molecular complexes that rotate in the plane and form the ellipticity of laser radiation polarization decreases.

Table 2.3 Statistical moments of the 1st–4th orders of magnitude, which characterize the distribution of the CDMP of the small-scale component of microscopic images of polycrystalline CSF films from groups 1 and 2

Statistical moments	Control group	Cause of death	
	Donors (n = 30)	Myocardial infarction (n = 30)	Alcohol poisoning (n = 30)
Average, M_1	0.19 ± 0.0026	0.13 ± 0.007	0.105 ± 0.006
P_1		$p \prec 0.05$	$p \prec 0.05$
P_2		$p \succ 0.05$	
Dispersion, M_2	0.13 ± 0.007	0.09 ± 0.005	0.07 ± 0.004
P_1		$p \prec 0.05$	$p \prec 0.05$
P_2		$p \succ 0.05$	
Skewness, M_3	0.79 ± 0.057	1.37 ± 0.074	1.03 ± 0.053
P_1		$p \prec 0.05$	$p \prec 0.05$
P_2		$p \succ 0.05$	
Kurtosis, M_4	0.64 ± 0.042	1.62 ± 0.084	1.12 ± 0.073
P_1		$p \prec 0.05$	$p \prec 0.05$
P_2		$p \prec 0.05$	

Table 2.4 Balanced accuracy in differentiating the cause of death by direct and spatial frequency mapping of the CDMP distributions of microscopic images of CSF polycrystalline films

Statistical moments	Average, M_1 (%)	Dispersion, M_2 (%)	Skewness, M_3 (%)	Kurtosis, M_4 (%)
Balanced accuracy of forward detection CDMP Ac	68	71	72	73
Differential large-scale detection of CDMP	69	72	73	72
Balanced accuracy of differential small-scale detection of CDMP	71	72	74	78

- A comparison of the histograms of the distribution of random values of the CDMP for the totality of points of the microscopic image shows that the range of scatter decreases by almost 3 times in the 6 h after death. At the same time, the number of extreme values of the CDMP grows by 2 times.

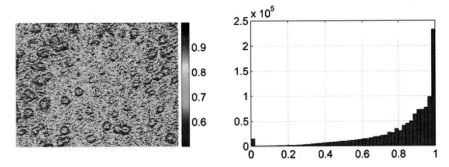

Fig. 2.1 Coordinate structure (left side) and distribution histogram (right side) of the random values of the CDMP of the microscopic image of PFCSD. AOD 3 h

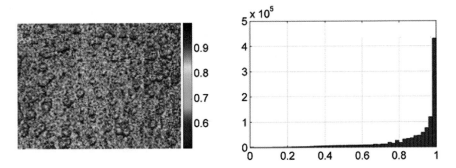

Fig. 2.2 Coordinate structure (left side) and distribution histogram (right side) of the random CDMP values of the set of points of the microscopic image of PFCSF. AOD 6 h

- Quantitatively, the optical manifestations of such degenerative changes in the polycrystalline structure of the CSF film are reflected in the rapid decrease in the 1st-order statistical moment (average) and the 2nd-order statistical moment (dispersion) of the distribution of random CDMP values of the set of points in the microscopic image of CSF polycrystalline films.
- Higher-order statistical moments (skewness and kurtosis) that characterize the distribution of the CDMP values should increase significantly.
- It has been established that the most sensitive and the most dynamically changing are the statistical moments of the 3rd and 4th orders, which characterize the skewness and kurtosis (peak sharpness) of the distribution of random values of the degree of mutual polarization of the set of points in the microscopic image of the polycrystalline CSF sample(s).

Figure 2.3 illustrates the diagrams of the temporal changes in the values of the 3rd and 4th order statistical moments, which are the most sensitive to postmortem degradation of the optically active structures in the CSF samples; they characterize the skewness and sharpness of the peak of the coordinate distributions of the CDMP

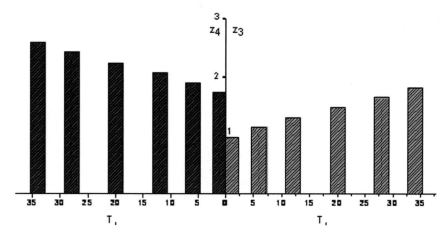

Fig. 2.3 Temporal dependences of the magnitude of the statistical moments of the 3rd–4th orders, which characterize the distribution of CDMP values of a microscopic image of polycrystalline CSF films of a human corpse

values of the set of points in the microscopic image of polycrystalline CSF films within 34 h after the onset of death (Table 2.5).

From the data on the temporary monitoring of changes in the magnitude of the statistical moments of higher orders that characterize the polarization-correlation structure of the distribution of the CDMP values in the plane of the microscopic images of polycrystalline CSF films, it follows that

- the range in the variation of the values of the statistical moment of the third order, which characterizes the skewness of the histogram of the distribution of random values of the CDMP, is 1.88 times;

Table 2.5 Temporal dependences of the statistical moments of the 3rd–4th orders of magnitude, which characterize the distribution of CDMP values of a set of points (pixels) of a digital microscopic image of polycrystalline CSF films of a human corpse

T, hours	1	6	12	20	28	34
Z_3	0.97 ± 0.031	1.14 ± 0.033	1.31 ± 0.035	1.48 ± 0.041	1.65 ± 0.047	1.82 ± 0.055
t	p	$t_{1;6} = 3.754$ $p \prec 0.001$	$t_{6;12} = 3.534$ $p \prec 0.001$	$t_{12;20} =$ 3.154 $p \prec 0.001$	$t_{20;28} =$ 2.726 $p \prec 0.001$	$t_{28;34} = 2.35$ $p \prec 0.05$
Z_4	1.74 ± 0.042	1.91 ± 0.049	2.08 ± 0.054	2.25 ± 0.066	2.44 ± 0.074	2.59 ± 0.086
t	p	$t_{1;6} = 2.634$ $p \prec 0.001$	$t_{6;12} = 2.331$ $p \prec 0.05$	$t_{12;20} =$ 1.993 $p \succ 0.05$	$t_{20;28} =$ 1.916 $p \succ 0.05$	$t_{28;34} =$ 1.322 $p \succ 0.05$

- the range in the variation of the values of the statistical moment of the fourth order, which characterizes the kurtosis (peak sharpness) of the histogram of the distribution of random values of the CDMP, is 1.81 times;
- based on the determination of the algorithm for the AOD, the following efficiency data were established for the polarization-correlation mapping method for the set of points of the microscopic images of polycrystalline CSF films: the determination interval of AOD—$T = 34$ h; the determination accuracy of AOD—$\Delta T = 35$ min.

Thus, it can be stated that when accounting for the coordinate features of the postmortem degenerative changes in the optical anisotropy parameters of the protein crystal networks by polarization-correlation mapping of CSF films, our approach can increase the establishment of the interval of AOD and improve its determination accuracy compared to polarimetric mapping techniques.

At the same time, the addition of the method for determining the coordinate distributions of the CDMP values by the technique of spatial-frequency filtering opens up additional possibilities in improving the indicators for determining the interval and the accuracy of AOD establishment.

2.3 Mapping of the Distributions of the Degree of Mutual Polarization of the Set of Points of Microscopic Images of Polycrystalline CSF Films with Large-Scale Spatial-Frequency Filtering

Next, we show the results from studying the temporal dynamics of the degenerative changes in the large-scale component of the birefringent polycrystalline protein networks by step-by-step scanning polarization-inhomogeneous microscopic images of CSF films under spatial-frequency filtering of coordinate distributions of the CDMP (Figs. 2.4, 2.5 and 2.6).

An analysis of the data of the two-dimensional polarization-correlation mapping of the distributions of the CDMP values obtained at the location of a Stocks-polarimeter with spatial-frequency filtering of polarization-inhomogeneous microscopic images of CSF polycrystalline films shows that degenerative changes in spatially oriented molecular networks at different times of observation after death occur sufficiently slowly (Figs. 2.4 and 2.5, left parts).

Quantitatively, this fact is illustrated by the minor changes in the histograms of the distribution of random values of the CDMP of the set of points (pixels) of digital microscopic images of large-scale molecular complexes of the CSF film (Figs. 2.4 and 2.5, right parts).

An analysis of the temporal degenerative transformation of the structure of polarization-correlation maps of the CDMP points of the microscopic images revealed that with an increase in the time of the onset of death, the optical birefringence of spatially oriented high molecular weight polycrystalline networks that

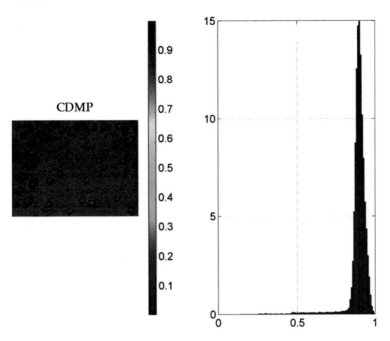

Fig. 2.4 The coordinate structure (left side) and the histogram of the distribution (right side) of the random values of the CDMP of the microscopic image of the large -scale component of the PFCSF. AOD 1 h

form elliptically polarized waves with different slopes of the plane of laser radiation polarization at points of the CSF film plane decreases sufficiently slowly compared with the data from the polarimetric mapping of large-scale components of the maps of the azimuth and ellipticity of the polarization.

Objectively, the growth of the determination interval of the AOD characterizes the comparison of histograms of the distribution of random values of CDMP of the large-scale components of the microscopic images of polycrystalline CSF films. As can be seen, the range and dispersion of the scatter of the random CDMP values do not significantly decrease (Figs. 2.4 and 2.5, right-hand sides) within 12 h after death.

Such differences in recording the rate of the degenerative cadaveric changes in the polycrystalline structure compared with direct two-dimensional polarization mapping of the azimuth and ellipticity can be because the CDMP value is due to the structure of closely spaced protein complexes. Significant degenerative changes for such PFCSF structures occur later after the onset of death. In the first place, the degradation of the optical anisotropy of PFCSF is manifested in ruinous high molecular weight complexes that have a high level of spatial orientation and, accordingly, significant birefringence, in medium and low molecular weight structures with a lower level of optical anisotropy.

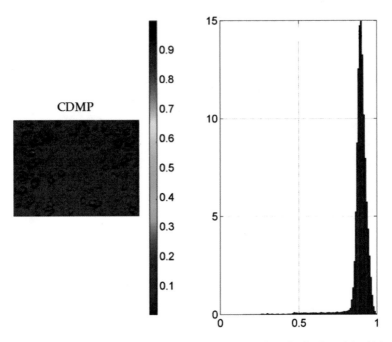

Fig. 2.5 The coordinate structure (left side) and the histogram of the distribution (right side) of the random values of the CDMP of the microscopic image of the large -scale component of the PFCSF. AOD 12 h

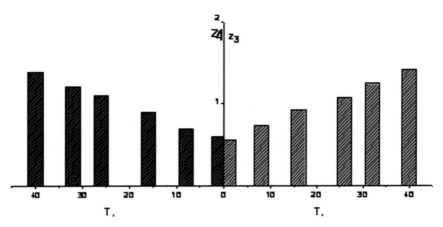

Fig. 2.6 Temporal dependences of the magnitude of the statistical moments of the 3rd–4th orders, which characterize the distribution of CDMP values of a microscopic image of the large-scale component of polycrystalline CSF films of a human corpse

Table 2.6 Temporal dependences of the statistical moments of the 3rd–4th orders of magnitude, which characterize the distribution of CDMP values of a set of points (pixels) of a digital microscopic image of polycrystalline CSF films of a human corpse

T, hours	1	8	16	26	32	40
Z_3	0.56 ± 0.016	0.73 ± 0.022	0.92 ± 0.025	1.07 ± 0.029	1.24 ± 0.04	1.41 ± 0.041
t	p	$t_{1;8} = 6.249$ $p \prec 0.001$	$t_{8;16} = 5.105$ $p \prec 0.001$	$t_{16;26} = 4.44$ $p \prec 0.001$	$t_{26;40} =$ 3.441 $p \prec 0.001$	$t_{40;48} =$ 2.968 $p \prec 0.001$
Z_4	0.39 ± 0.018	0.69 ± 0.021	0.89 ± 0.023	1.102 ± 0.031	1.21 ± 0.038	1.28 ± 0.039
t	p	$t_{1;8} = 6.126$ $p \prec 0.001$	$t_{8;16} = 5.083$ $p \prec 0.001$	$t_{16;26} = 4.44$ $p \prec 0.001$	$t_{26;40} =$ 3.423 $p \prec 0.001$	$t_{40;48} =$ 2.941 $p \prec 0.001$

Quantitatively, this scenario within the statistical approach to the analysis of polarization-correlation maps results in a slight decrease in the average and variance, which characterize the distribution of random values of the complex degree of mutual polarization of the set of points of microscopic images of polycrystalline films of CSF with an increased time in the observation after the onset of death.

Higher-order statistical moments (skewness and kurtosis) of such polarization-correlation distributions of CDMP values, in contrast, should increase significantly.

Therefore, in the process of temporary monitoring of postmortem changes in the polarization-correlation maps of high molecular weight protein components of polycrystalline films of CSF, the statistical moments of the 3rd and 4th orders, which characterize the skewness and kurtosis (peak sharpness) of the distribution of random values of the complex degree of mutual polarization of the set of points of digital microscopic images.

Table 2.6 presents the results of the calculations of the statistical moments of the 3rd–4th orders, which characterize the coordinate distributions of the CDMP values of the large-scale component of microscopic images of polycrystalline networks of albumin and fibrin from CSF films within 40 h after death.

An analysis was performed of the temporal dynamics of changes in the statistical structure of the distributions of the CDMP values of microscopic images of large-scale protein polycrystalline networks of CSF films obtained at the location of a Stocks-polarimeter with spatial-frequency filtering of polarization-inhomogeneous laser images. The results revealed the following:

- the range of variation in the values of the statistical moment of the 3rd order, which characterizes the skewness of the distribution of the values of the CDMP, is 2.45 times during the 48 h of observation after death;
- the range of changes in the values of the statistical moment of the 4th order, which characterizes the kurtosis of the distribution of the CDMP values, is 1.4 times during the 48 h of observation after death;
- a comparative analysis with the data from polarization-correlation mapping of classical microscopic images revealed an increase in the AOD determination

interval due to a slower drop in the values of the statistical moments of the 3rd and 4th orders, which was $T = 48$ h;
- the accuracy of determining the AOD for the methods of polarization and polarization-correlation mapping is quite high, $\Delta T = 1$ год.

2.4 Mapping the Distributions of the Degree of Mutual Polarization of the Set of Points of Microscopic Images of Polycrystalline CSF Films with Small-Scale Spatial-Frequency Filtering

Using the approach of studying the capabilities of the method of polarization-correlation mapping of polarization-inhomogeneous microscopic images of CSF films to determine the interval and accuracy of the AOD establishment by applying high-frequency spatial-frequency filtering, provides a direct statistical analysis of the optical manifestations of small-scale molecular structures, the results are shown in a Figs. 2.7 and 2.8.

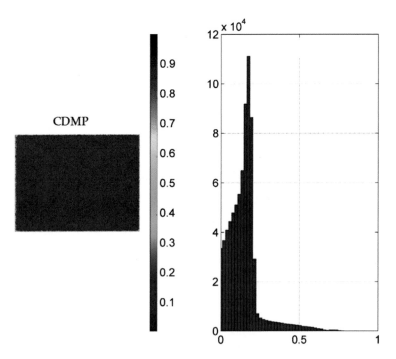

Fig. 2.7 The coordinate structure (left side) and the histogram of the distribution (right side) of the random values of the CDMP of the microscopic image of the small-scale component of the PFCSF. AOD 1 h

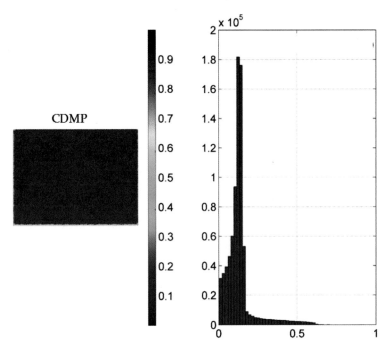

Fig. 2.8 The coordinate structure (left side) and the histogram of the distribution (right side) of the random values of the CDMP of the microscopic image of the small-scale component of the PFCSF. AOD 12 h

From a fundamental point of view, the obtained coordinate distributions of the values of the complex degree of mutual polarization of the set of points of the microscopic image carry more detailed information about the postmortem biochemical changes in CSF at the concentration or primary level of its polycrystalline structure. Therefore, one should expect an increase in the sensitivity and accuracy of determining the AOD in the short-term after death.

Experimental results confirm the foregoing predictive score. As seen from a comparative analysis of the results of direct polarization-correlation mapping and data of polarimetric mapping with low-frequency spatial-frequency filtering of microscopic images of polycrystalline CSF films, the optical-physical technique for correlating the studies of degenerative changes in small-scale polycrystalline networks at different times of observation after death is very efficient (Figs. 2.7 and 2.8, left parts).

As seen, the polarization-correlation maps of optically active molecular compounds, which form the coordinate distributions of the values of the complex degree of mutual polarization, are a set of small-scale zones (polarization domains), the number of which decreases significantly with increasing AOD.

This fact can be attributed to the fact that degenerative biochemical changes begin with a decrease in optically anisotropic domains due to the destruction of high molecular weight compounds and the formation of low molecular weight optically isotropic (without changing polarization) molecular complexes. Quantitatively, this scenario is illustrated by significant changes (a 10-fold decrease) in the zero values of the CDMP, which characterize the maximum decorrelation of the polarization states at closely located points of the CSF film. For this reason, the histograms of the distribution of random values of the complex degree of mutual polarization change very quickly in the sense of transformation of the values of skewness and kurtosis (Figs. 2.7 and 2.8, right-hand sides) for the first 6 h after death.

Table 2.3 presents the results of the calculations of the statistical moments of the 3rd and 4th orders of magnitude, which characterize the coordinate distributions of the CDMP values of the small-scale component of microscopic images of polycrystalline networks of globulin and glucose in CSF films within 16 h after death.

Figure 2.9 illustrates the diagrams of the temporal changes in the values of the 3rd and 4th order statistical moments, which characterize the polarization-correlation maps of such samples; these changes are the most sensitive to post-mortem degradation of optically active molecular complexes in polycrystalline CSF films (Table 2.7).

An analysis of the temporal dynamics of the transformation of the statistical structure of the coordinate distributions of the values of the CDMP of microscopic images of the small-scale component of polycrystalline networks of CSF films revealed that such changes are most clearly and dynamically manifested in variations in the values of the higher-order statistical moments. In particular, over the period of 3 h after death, the skewness of the histograms of the distribution of the CDMP increased by 1.79 times, and the kurtosis of the distribution grew 2.04 times.

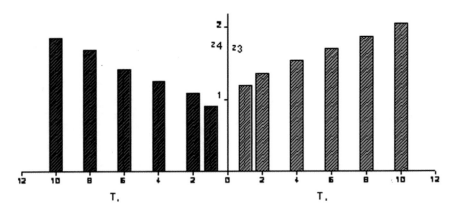

Fig. 2.9 Temporal dependences of the magnitudes of the statistical moments of the 3rd–4th orders, which characterize the distributions of the CDMP values of a microscopic image of the small-scale component of polycrystalline CSF films

Table 2.7 Time dependences of the magnitudes of the statistical moments of the 3rd and 4th orders, which characterize the distribution of the CDMP values of the microscopic image of the small-scale component of polycrystalline CSF films of a human corpse

T, hours	1	2	4	6	8	10
Z_3	1.19 ± 0.032	1.36 ± 0.037	1.53 ± 0.041	1.7 ± 0.046	1.87 ± 0.049	2.04 ± 0.06
t	p	$t_{1;2} = 3.475$ $p \prec 0.001$	$t_{2;4} = 3.078$ $p \prec 0.001$	$t_{4;6} = 2.759$ $p \prec 0.001$	$t_{6;8} = 2.529$ $p \prec 0.05$	$t_{8;10} = 2.195$ $p \prec 0.05$
Z_4	0.91 ± 0.035	1.08 ± 0.031	1.25 ± 0.036	1.42 ± 0.04	1.69 ± 0.048	1.86 ± 0.051
t	p	$t_{1;2} = 3.636$ $p \prec 0.001$	$t_{2;4} = 3.578$ $p \prec 0.001$	$t_{4;6} = 3.159$ $p \prec 0.001$	$t_{6;8} = 3.321$ $p \prec 0.001$	$t_{8;10} = 2.427$ $p \prec 0.001$

A comparative analysis with the data from polarization-correlation mapping of classical and large-scale microscopic images revealed a decrease in the AOD determination interval due to the most rapid increase in the values of the statistical moments of the 3rd and 4th orders, at $T = 10$ h. At the same time, the maximum accuracy of AOD determination is reached, $\Delta T = 20$ min.

2.5 Intervals and Accuracy of AOD Measurements by Two-Dimensional Mapping of Distributions of Values of the Complex Degree of Mutual Polarization of Microscopic Images of CSF Polycrystalline Films

This paragraph provides comparative data on the intervals and the accuracy of determining the AOD by the method of step-by-step scanning of polarization-inhomogeneous classical and spatial-frequency filtered microscopic images of polycrystalline CSF films, as in Table 2.8.

An analysis of the parameters (interval and accuracy) of the AOD determination by polarization-correlation mapping of the distributions of the complex degree of mutual polarization of the microscopic images of CSF polycrystalline films using the classical two-dimensional polarimetry method and spatial-frequency filtering showed the maximum accuracy of the AOD determination among all of the known methods of laser polarimetry.

As in the case of direct polarization mapping, a set of polarization-correlation methods for determining the CDMP distributions of a set of points of polarization-inhomogeneous images of the CSF polycrystalline films of donors (group 0) and those who died as a result of myocardial infarction (group 1) and severe alcohol poisoning (group 2) was established, with the following properties:

1. provides the ability to differentiate samples of CSF polycrystalline films of deceased and donors according to the calculation of the statistical moments of the 4th order, with a confidence interval $p \prec 0.05$ (Tables 2.1, 2.2, 2.3 and 2.4)

Table 2.8 Intervals and accuracy of AOD determination by mapping CDMP values of microscopic images of polycrystalline CSF films

Parameters	AOD detection interval, T	Statistical parameters	AOD determination accuracy, ΔT
CDMP distribution of the microscopic image	30 h	$Z_3 = 0.97 - 1.82$ $Z_4 = 1.74 - 2.59$	35 min
CDMP distribution of the large-scale component of the microscopic image	48 h	$Z_3 = 0.56 - 1.41$ $Z_4 = 0.39 - 1.28$	1 h
CDMP distribution of the small-scale component of the microscopic image	10 h	$Z_3 = 1.19 - 2.04$ $Z_4 = 0.91 - 1.86$	25 min

2. found a statistical moment of the fourth order that is reliable for the case of severe alcohol poisoning ($p \prec 0.05$), the calculation of which provides a satisfactory level of balanced accuracy $Ac(M_4) = 79\%$ (Table 2.4) differentiation of the cause of death

3. for all of the other cases, it does not provide statistically significant differentiation of polarization-correlation maps of CDMP microscopic images of CSF polycrystalline films due to the following:

   (a) confidence interval for all of the statistical moments, 1st–4th orders, is $p \succ 0.05$;
   (b) the value of the balanced accuracy of differentiation of various causes of death using these methods does not reach a satisfactory level: direct mapping of CDMP (Table 5.7, $Ac \leq 64\%$); and mapping large (Table 5.7, $Ac \leq 66\%$) and small-scale (Table 5.7, $Ac \leq 72\%$) components of the CDMP maps.

Chapter 3
Studies of the Forensic Effectiveness of the Determination of the Time of Death Onset Based on Laser-Induced Fluorescence of Polycrystalline Films of Cerebrospinal Fluid

This part contains the results of a study of the temporal dynamics of postmortem changes in the biochemical molecular structure of CSF by determining the statistical moments of the 1st–4th orders, which characterize the coordinate distributions of the values of the intensity and azimuth of polarization of the laser-induced fluorescence of samples of polycrystalline films.

The coordinate distributions of the intensities and polarization azimuths of laser-induced autofluorescence of polycrystalline CSF films were measured using band-pass interference filters with the following spectral transmission maxima: $\lambda_1 = 0.45\,\text{mkm}$, $\lambda_2 = 0.55\,\text{mkm}$ and $\lambda_3 = 0.63\,\text{mkm}$.

By summarizing the time dependences of the most sensitive statistical moments of the 1st–4th orders, which characterize spectrally selective fluorescence images and polarization maps of molecular polycrystalline networks of CSF films within statistically significant groups of samples, the intervals and the accuracy of establishing the time of the onset of death are determined.

3.1 Investigation of the Temporal Dynamics of Post-mortem Changes in the Statistical Structure of Spectral-Selective Intensity Distributions of Laser-Induced Fluorescence of Polycrystalline Molecular Films of CSF

Optical-physical fluorescence methods are widely and very effectively used in the medical diagnosis of various diseases.

© The Author(s), under exclusive licence to Springer Nature Singapore Pte Ltd. 2021

M. S. Harazdyuk et al., *Correlation and Autofluorescence Microscopy in Forensics Medicine: Time of Death Detection Using Polycrystalline Cerebrospinal Fluid Films*, SpringerBriefs in Physics, https://doi.org/10.1007/978-981-16-0197-2_3

Such methods are based on the use of the phenomenon of fluorescence—the secondary radiation of various molecular structures that occurs under the influence of short-wavelength optical radiation to biological tissue or fluid.

The following main stages of fluorescence diagnostics are distinguished:

1. The spectral stage—the assessment of fluorescence by the color of the radiation—light blue, blue, green, yellow and red.
2. Spectrophotometry—measurement of the fluorescence spectrum—distribution of the fluorescence intensity over wavelengths.
3. Analytical—comparison of fluorescence spectra upon excitation of a biological object by radiation with different wavelengths.

All fluorescence methods are based on the phenomenon of absorption of optical radiation by various molecular compounds. For this reason, the molecules are excited and emit fluorescent light in the longer wavelength region of the spectrum of optical radiation.

In medical diagnostics, using absorption spectral techniques, the absorption spectra of biological substances at various wavelengths are measured.

In our work, we used a technique that is based on the excitation of the intrinsic fluorescence of biological molecules by laser radiation: laser-induced fluorescence. This method allows us to combine the study of the distribution of intrinsic fluorescence intensities in various spectral regions, as well as the measurement of the distributions of its polarization parameters, more specifically, polarization azimuths that characterize optically active chiral molecules.

The following groups of fluorophores are known for occurring in biological tissues and fluids, and they emit the most intensely in various spectral regions:

- The short wavelength "light blue–blue" region $\Delta\lambda = 0.46\,\text{mkm}-0.48\,\text{mkm}$—proteins, nicotinamide-adenine-dinucleotide (NADR).
- The medium wavelength "green–yellow" region $\Delta\lambda = 0.52\,\text{mkm}-0.54\,\text{mkm}$—flavins and folic acids.
- The long-wavelength "red" region $\Delta\lambda = 0.58\,\text{mkm}-0.66\,\text{mkm}$—porphyrins.

When mapping the intensity distribution of laser-induced fluorescence, a bandpass filter with a maximum transmission was used, $\lambda_1 = 0.45\,\text{mkm}$.

Samples of CSF films on optically homogeneous glass were made using the procedure described above and were used as objects of this study.

Experimental studies of the temporal dynamics of the postmortem changes in the biochemical structure of proteins and NADR in CSF films by measuring the coordinate distribution of the intensity of laser-induced intrinsic fluorescence were conducted according to the following algorithm:

1. For each PFCSF sample in the optical arrangement of the Stokes-polarimeter, the samples were irradiated with the radiation of a blue semiconductor laser, LSR405ML-LSR-PS-II, with a wavelength $\lambda = 0.405\,\text{mkm}$ and power $W = 50\,\text{mBm}$.

2. Using a digital camera, we measured spectrally selective two-dimensional distributions of intrinsic fluorescence intensities of CSF polycrystalline films.
3. For each two-dimensional distribution of the intensity values of the laser-induced intrinsic fluorescence of PFCSF, the values of the statistical moments of the 1st–4th orders were calculated.
4. Statistical processing of the measured set of values of the statistical moments 1st–4th orders, which characterize the distribution of the intensity values of the laser-induced intrinsic fluorescence of PFCSF, was conducted within a representative sample of samples until the value of the confidence interval was reached, $p \prec 0.01$.
5. The time dependences of changes in the magnitude of the statistical moments that are the most sensitive to changes in the structure of PFCSF fluorescence maps were constructed over a dynamic interval—the "variation—stabilization" of the values.
6. The interval and accuracy of AOD establishment were determined.

3.2 Differentiation of the Cause of Death Based on Autofluorescence Microscopy of Polycrystalline Films of Cerebrospinal Fluid

This part of the work presents the results of a statistical analysis of the distribution of the intensities of fluorescence in the cerebrospinal fluid films (**"heart failure—severe alcohol poisoning"**) in different parts of the spectrum ("blue"—Table 3.9; "green-yellow"—Table 3.10; "red"—Table 3.11) spectrum.

3.2.1 "Blue" Region of the Spectrum

See Table 3.1.

3.2.2 "Green-Yellow" Region of the Spectrum

See Table 3.2.

3.2.3 "Red" Region of the Spectrum

Autofluorescence polarimetry results of the coordinate distributions of polarization azimuths of microscopic images of molecular fluorophores of polycrystalline films

Table 3.1 Statistical moments of the 1st–4th orders, which characterize the intensity distribution of microscopic images of autofluorescence of polycrystalline films of cerebrospinal fluid

Statistical moments	Control group	Cause of death	
	Donors (n = 30)	Myocardial infarction (n = 30)	Donors (n = 30)
Average, M_1	0.41 ± 0.026	0.22 ± 0.017	0.24 ± 0.019
P_1		$p < 0.05$	$p < 0.05$
P_2		$p > 0.05$	
Dispersion, M_2	0.13 ± 0.007	0.08 ± 0.004	0.09 ± 0.006
P_1		$p < 0.05$	$p < 0.05$
P_2		$p > 0.05$	
Skewness, M_3	0.38 ± 0.017	0.91 ± 0.054	0.63 ± 0.031
P_1		$p < 0.05$	$p < 0.05$
P_2		$p < 0.05$	
Kurtosis, M_4	0.43 ± 0.024	1.25 ± 0.064	0.81 ± 0.043
P_1		$p < 0.05$	$p < 0.05$
P_2		$p < 0.05$	

Table 3.2 Statistical moments the 1st–4th orders, which characterize the intensity distribution of the autofluorescence of polycrystalline films of cerebrospinal fluid

Statistical moments	Control group	Cause of death	
	Donors (n = 30)	Myocardial infarction (n = 30)	Donors (n = 30)
Average, M_1	0.44 ± 0.026	0.19 ± 0.011	0.27 ± 0.016
P_1		$p < 0.05$	$p < 0.05$
P_2		$p < 0.05$	
Dispersion, M_2	0.19 ± 0.011	0.11 ± 0.006	0.15 ± 0.008
P_1		$p < 0.05$	$p < 0.05$
P_2		$p < 0.05$	
Skewness, M_3	0.21 ± 0.013	0.92 ± 0.047	0.61 ± 0.031
P_1		$p < 0.05$	$p < 0.05$
P_2		$p < 0.05$	
Kurtosis, M_4	0.19 ± 0.012	0.55 ± 0.029	0.36 ± 0.017
P_1		$p < 0.05$	$p < 0.05$
P_2		$p < 0.05$	

of cerebrospinal fluid of all three studied groups for different parts of the spectrum are shown (Table 3.3).

Table 3.3 Statistical moments of the 1st–4th orders, which characterize the intensity distribution of the autofluorescence of polycrystalline films of cerebrospinal fluid

Statistical moments	Control group	Cause of death	
	Donors (n = 30)	Myocardial infarction (n = 30)	Donors (n = 30)
Average, M_1	0.39 ± 0.021	0.11 ± 0.006	0.19 ± 0.011
P_1		$p \prec 0.05$	$p \prec 0.05$
P_2		$p \prec 0.05$	
Dispersion, M_2	0.19 ± 0.012	0.11 ± 0.006	0.14 ± 0.008
P_1		$p \prec 0.05$	$p \prec 0.05$
P_2		$p \prec 0.05$	
Skewness, M_3	0.68 ± 0.037	0.99 ± 0.053	1.31 ± 0.071
P_1		$p \prec 0.05$	$p \prec 0.05$
P_2		$p \prec 0.05$	
Kurtosis, M_4	0.83 ± 0.044	1.53 ± 0.084	1.09 ± 0.063
P_1		$p \prec 0.05$	$p \prec 0.05$
P_2		$p \prec 0.05$	

3.2.4 "Blue" Region of the Spectrum

See Table 3.4.

Table 3.4 Statistical moments of the 1st–4th orders, which characterize the distribution of azimuths of polarization of the autofluorescence of CSF polycrystalline films

Statistical moments	Control group	Cause of death	
	Donors (n = 30)	Myocardial infarction (n = 30)	Donors (n = 30)
Average, M_1	0.11 ± 0.006	0.06 ± 0.004	0.05 ± 0.003
P_1		$p \prec 0.05$	$p \prec 0.05$
P_2		$p \succ 0.05$	
Dispersion, M_2	0.09 ± 0.005	0.05 ± 0.003	0.06 ± 0.004
P_1		$p \prec 0.05$	$p \prec 0.05$
P_2		$p \succ 0.05$	
Skewness, M_3	0.63 ± 0.041	1.33 ± 0.072	0.94 ± 0.052
P_1		$p \prec 0.05$	$p \prec 0.05$
P_2		$p \prec 0.05$	
Kurtosis, M_4	0.79 ± 0.043	1.49 ± 0.084	1.05 ± 0.053
P_1		$p \prec 0.05$	$p \prec 0.05$
P_2		$p \prec 0.05$	

Table 3.5 Statistical moments of the 1st–4th orders, which characterize the distribution of the azimuths of polarization of the autofluorescence of CSF polycrystalline films

Statistical moments	Control group	Cause of death	
	Donors (n = 30)	Myocardial infarction (n = 30)	Donors (n = 30)
Average, M_1	0.12 ± 0.007	0.08 ± 0.005	0.046 ± 0.003
P_1		$p \prec 0.05$	$p \prec 0.05$
P_2		$p \prec 0.05$	
Dispersion, M_2	0.09 ± 0.005	0.06 ± 0.004	0.04 ± 0.002
P_1		$p \prec 0.05$	$p \prec 0.05$
P_2		$p \prec 0.05$	
Skewness, M_3	0.48 ± 0.029	1.08 ± 0.058	0.73 ± 0.039
P_1		$p \prec 0.05$	$p \prec 0.05$
P_2		$p \prec 0.05$	
Kurtosis, M_4	0.81 ± 0.044	1.43 ± 0.078	1.06 ± 0.059
P_1		$p \prec 0.05$	$p \prec 0.05$
P_2		$p \prec 0.05$	

3.2.5 "Green-Yellow" Region of the Spectrum

Analysis of the data obtained by the method of mapping the laser-induced fluorescence revealed the following (Tables 3.5 and 3.6):

- In the "shortwave" region of the spectrum, there is a clear dependence of the fluorescence of the proteins and the PFCSF of the NADR compounds on the time of the onset of death. This fact is indicated by a change in the magnitude of the maximum intensity and the range of scatter in the values in the histograms of the distribution of this parameter (Figs. 3.1 and 3.2, right parts).
- With an increase in the observation time after death due to degenerative changes in CSF films at the molecular level, the concentration of proteins and NADR decreases rapidly.
- A comparison of the histograms of the distribution of random values of intrinsic fluorescence (Figs. 3.1 and 3.2, right-hand sides) shows that the extreme value of the fluorescence intensity decreases by almost 2.5 times: $l_{max}^{\Phi}(T = 1\,\mathrm{h}) = 0.36 \rightarrow l_{max}^{\Phi}(T = 6\,\mathrm{h}) = 0.14$ 6 h after death (Fig. 3.3).
- Quantitatively, the optical manifestations of such degenerative changes in the molecular structure of the CSF film are reflected in a decrease in the average and a dispersion of the distribution of random values of the azimuths of polarization. The statistical moments of higher orders (skewness and kurtosis), in contrast, grow.
- It has been established that the most sensitive and, thus, the most dynamically changing entities are the statistical moments of the 1st, 3rd and 4th orders, which

Table 3.6 Balanced accuracy in differentiating the causes of death by spectral-selective mapping of the distributions of intensity and azimuths of microscopic images of the autofluorescence of polycrystalline films of cerebrospinal fluid

Statistical moments	Average, M_1	Dispersion, M_2	Skewness, M_3	Kurtosis, M_4
Balanced photometry accuracy in the "blue" region of the fluorescence spectrum, Ac	74% (66%)	72% (64%)	74% (66%)	80% (68%)
Balanced photometry accuracy in the "green-yellow" region of the fluorescence spectrum, Ac	76% (68%)	80% (72%)	82% (72%)	82% (74%)
Balanced photometry accuracy in the "red" region of the fluorescence spectrum, Ac	78% (70%)	82% (72%)	88% (74%)	90% (76%)
Balanced polarimetry accuracy in the "blue" region of the fluorescence spectrum, Ac	76% (68%)	74% (64%)	78% (68%)	82% (70%)
Balanced polarimetry accuracy in the "green-yellow" region of the fluorescence spectrum, Ac	78% (66%)	80% (68%)	82% (72%)	84% (72%)
Balanced polarimetry accuracy in the "red" region of the fluorescence spectrum, Ac	82% (70%)	84% (70%)	90% (76%)	92% (78%)

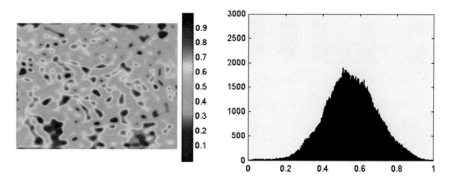

Fig. 3.1 Coordinate structure (left side) and distribution histogram (right side) of random values of the intensity of laser-induced intrinsic fluorescence of CSF polycrystalline films. AOD 3 h

characterize the average, skewness and kurtosis of the distribution of random values of fluorescence intensity.

Tables 3.7 and 3.8 show the results of calculations of the statistical moments of the 1st, 3rd, and 4th orders, which characterize the coordinate distributions of the values of the intensity of laser-induced fluorescence of protein molecules and NADR of the CSF polycrystalline films within 24 h after death.

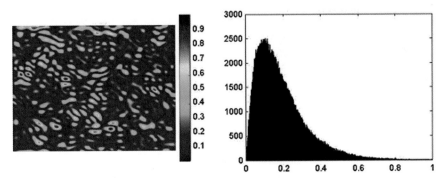

Fig. 3.2 Coordinate structure (left side) and distribution histogram (right side) of random values of the intensity of laser-induced intrinsic fluorescence of CSF polycrystalline films. AOD 6 h

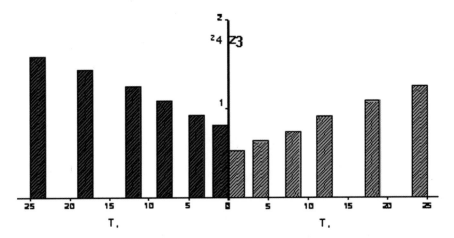

Fig. 3.3 Temporal dependences of the statistical moments of the 3rd–4th orders, which characterize the distribution of the fluorescence intensities of protein molecules and NADR in polycrystalline CSF films

From a statistical analysis of the dynamics of temporary changes in the coordinate distributions of maps of laser-induced fluorescence of protein molecules and nicotinamide nucleotide NADR of CSF polycrystalline films, we found the following:

- the range of variation in the values of the statistical moment of the first order, which characterizes the average distribution of the values of the intrinsic fluorescence intensity of the CSF samples in the "blue" region of the spectrum, is 2.14 times;
- the range of variation in the values of the statistical moment of the third order, which characterizes the skewness of the distribution of the intrinsic fluorescence intensity values of the CSF samples in the "blue" region of the spectrum, is 2.87 times;

Table 3.7 Temporal dependences of the magnitudes of the statistical moments of the 1st, 3rd, and 4th orders, which characterize the distribution of the values of the autofluorescence intensity of the protein molecules and NADR in the CSF films

T, hours	1	4	8	12	18	24
Z_1	0.38 ± 0.016	0.32 ± 0.019	0.27 ± 0.023	0.22 ± 0.029	0.17 ± 0.033	0.16 ± 0.037
t	p	$t_{1;4} = 2.416$ $p \prec 0.001$	$t_{4;8} = 1.676$ $p \succ 0.05$	$t_{8;12} =$ 1.351 $p \succ 0.05$	$t_{12;18} =$ 1.138 $p \succ 0.05$	$t_{18;24} =$ 0.202 $p \succ 0.05$
Z_3	0.52 ± 0.016	0.63 ± 0.019	0.74 ± 0.023	0.91 ± 0.029	1.08 ± 0.033	1.45 ± 0.047
t	p	$t_{1;4} = 4.428$ $p \prec 0.001$	$t_{4;8} = 3.687$ $p \prec 0.001$	$t_{8;12} =$ 4.593 $p \prec 0.001$	$t_{12;18} = 3.87$ $p \prec 0.001$	$t_{18;24} =$ 6.443 $p \prec 0.001$
Z_4	0.81 ± 0.024	0.92 ± 0.028	1.08 ± 0.031	1.25 ± 0.035	1.44 ± 0.041	1.79 ± 0.069
t	p	$t_{1;4} = 2.983$ $p \prec 0.001$	$t_{4;8} = 3.83$ $p \prec 0.001$	$t_{8;12} =$ 3.636 $p \prec 0.001$	$t_{12;18} =$ 3.525 $p \prec 0.001$	$t_{18;24} =$ 4.361 $p \prec 0.01$

Table 3.8 Temporal dependences of the values of the statistical moments of the 1st, 3rd, and 4th orders, which characterize the distribution of the values of the fluorescence intensity of flavin and folic acid molecules in CSF polycrystalline films of a corpse

T, hours	1	4	10	16	22	28
Z_1	0.48 ± 0.021	0.31 ± 0.013	0.27 ± 0.012	0.22 ± 0.009	0.17 ± 0.008	0.11 ± 0.006
t	p	$t_{1;4} = 3.526$ $p \prec 0.001$	$t_{4;10} =$ 2.261 $p \prec 0.005$	$t_{10;16} =$ 3.333 $p \prec 0.001$	$t_{16;22} =$ 4.152 $p \prec 0.001$	$t_{22;28} = 6$ $p \prec 0.001$
Z_3	0.56 ± 0.015	0.81 ± 0.022	0.97 ± 0.028	1.14 ± 0.033	1.31 ± 0.037	1.58 ± 0.042
t	p	$t_{1;4} = 9.389$ $p \prec 0.001$	$t_{4;10} =$ 4.493 $p \prec 0.001$	$t_{10;16} =$ 3.928 $p \prec 0.001$	$t_{16;22} =$ 3.429 $p \prec 0.001$	$t_{22;28} =$ 3.037 $p \prec 0.001$
Z_4	0.33 ± 0.008	0.51 ± 0.015	0.68 ± 0.02	0.85 ± 0.024	1.02 ± 0.033	1.19 ± 0.034
t	p	$t_{1;4} =$ 10.588 $p \prec 0.001$	$t_{4;10} = 6.8$ $p \prec 0.001$	$t_{10;16} =$ 5.441 $p \prec 0.001$	$t_{16;22} =$ 4.166 $p \prec 0.001$	$t_{22;28} =$ 3.588 $p \prec 0.001$

- the range of variation in the values of the statistical moment of the fourth order, which characterizes the sharpness of the peak distribution of the values of the intrinsic fluorescence intensity of the CSF samples in the blue region of the spectrum, is 1.87 times;

- Based on the AOD determination algorithm, the following data were established: AOD determination interval: $T = 24$ h; accuracy of determination of AOD: $\Delta T = 25$ min.

When mapping the distributions of the values of the intensity of laser-induced fluorescence, we used a bandpass filter with a maximum transmission $\lambda_2 = 0.55$ mkm, which made it possible to detect intrinsic reradiation of flavins and folic acid.

We performed an analysis of the two-dimensional mapping data obtained at the location of the Stokes-polarimeter with spectrally selective filtering of microscopic images of the intrinsic fluorescence of CSF polycrystalline films; the results showed a fairly rapid decrease in the intensity of the secondary radiation of the biochemical compounds in the mid-wave spectral range over a short (6 h) time period after death (Figs. 3.4 and 3.5, left parts).

Quantitatively, this process is illustrated by significant changes (3.5 times) in the average intensity value ($l_{max}^{\Phi}(T = 1\,\text{h}) = 0.35 \rightarrow l_{max}^{\Phi}(T = 6\,\text{h}) = 0.095$) of the

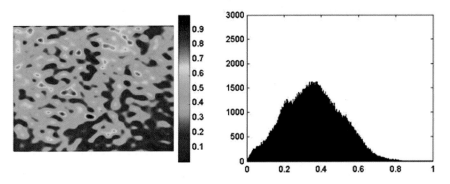

Fig. 3.4 Coordinate structure and histogram of the distribution of fluorescence intensity ($\lambda_2 = 0.55$ mkm) of PFCSF. AOD 1 h

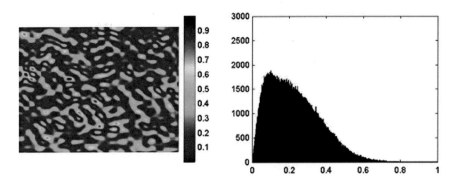

Fig. 3.5 Coordinate structure (left side) and distribution histogram (right side) of random values of the intensity of laser-induced fluorescence of PFCSF in the mid-wavelength region of the optical radiation spectrum. AOD 6 h

fluorescence of flavins and folic acids of CSF films in histograms of the distribution of this parameter (Figs. 3.4 and 3.5, right parts).

As a result of temporary monitoring of postmortem changes in microscopic images of laser-induced fluorescence of molecular compounds of CSF polycrystalline films in the green-yellow region of the spectrum of optical radiation, the statistical moments of the 1st (fastest growing, 2.6 times), 3rd (largest, 3.6 times) and 4th (growing 3.37 times) orders characterize the distribution of the intensity values of the secondary re-emission of flavins and folic acid.

Table 3.2 and Fig. 3.6 present the results of calculations of the statistical moments of the 1st, 3rd and 4th orders, which characterize degenerative changes in the coordinate distributions of the intensities of microscopic images of intrinsic fluorescence of polycrystalline CSF films within 28 h after death.

A comparative analysis of the temporal changes in the structure of fluorescence emission maps in the short and medium wavelength spectral ranges of the optical radiation revealed an increase in the AOD determination interval for the longer wavelength range of $T = 28$ h. In this case, the accuracy of the AOD determination decreases somewhat to $\Delta T = 35$ min.

When mapping the distributions of the values of the intensity of laser-induced fluorescence, we used a bandpass filter that had a maximum transmission $\lambda_3 = 0.63$ mkm, which made it possible to detect intrinsic reradiation of porphyrin molecules.

The results of the study of the possibilities of determining AOD by applying two-dimensional mapping of the intensity distributions of laser-induced fluorescence of CSF molecular polycrystalline films in the red region of the spectrum are shown in Figs. 3.7 and 3.8. Quantitatively, this fact is illustrated by significant changes (a 4-fold decrease) in the average scatter of the random values of the intensity of laser-induced fluorescence of porphyrins of CSF films of a human corpse (Figs. 3.7 and 3.8, right-hand sides) in the shortest time period (3:00) after death.

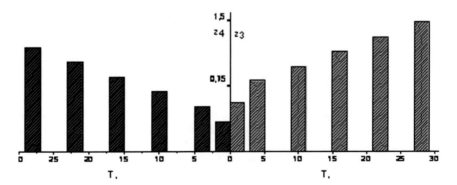

Fig. 3.6 Temporal dependences of the statistical moments of the 3rd and 4th orders, which characterize the distribution of the intensities of the laser-induced intrinsic fluorescence of CSF polycrystalline films of a human corpse in the green-yellow region of the spectrum of optical radiation

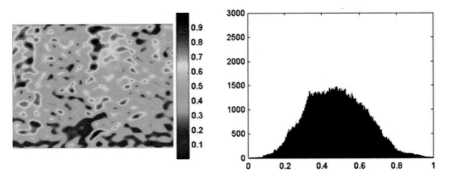

Fig. 3.7 Coordinate structure (left side) and distribution histogram (right side) of the intensity values of the microscopic image of the fluorescence of the PFCSF of a corpse. AOD 1 h

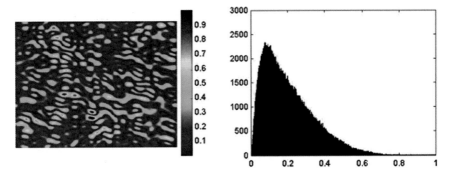

Fig. 3.8 Coordinate structure (left side) and distribution histogram (right side) of the intensity values of the microscopic image of the fluorescence of the PFCSF of a corpse. AOD 3 h

Table 3.9 presents the results of calculations of the magnitude of the most sensitive statistical moments that characterize the coordinate distribution of intrinsic fluorescence intensities of porphyrins of polycrystalline networks of CSF films within 14 h after death.

It has been established that the most distinct and dynamic temporal changes in the coordinate structure of laser-induced fluorescence maps of porphyrins of polycrystalline networks of CSF films are reflected in changes in the values of the 1st-order statistical moment, which characterizes the average value of the fluorescence intensity in the red spectral region: the range is 2.34 times. The range of the variation in the values of the statistical moment of the third order, which characterizes the skewness of the histogram of the distribution of random values of fluorescence intensity, is 3.34 times. For kurtosis, the range of values is 2.27 times.

The maximum decrease in the AOD determination interval ($T = 14$ h) for this optical-physical technique was revealed. In this case, a maximum accuracy of AOD determination of $\Delta T = 15$ min is achieved, as shown in Table 3.10.

Table 3.9 Temporal dependences of the statistical moments of the 1st, 3rd, and 4th orders, which characterize the distribution of the intensities of laser-induced fluorescence of porphyrin molecules of CSF polycrystalline films of a corpse

T, hours	1	3	5	7	10	14
Z_1	0.54 ± 0.019	0.45 ± 0.016	0.38 ± 0.015	0.33 ± 0.013	0.28 ± 0.012	0.23 ± 0.009
t	p	$t_{1;3} = 3.623$ $p \prec 0.001$	$t_{3;5} = 3.191$ $p \prec 0.001$	$t_{5;7} = 2.518$ $p \prec 0.001$	$t_{7;10} =$ 2.826 $p \prec 0.001$	$t_{10;14} =$ 3.333 $p \prec 0.001$
Z_3	0.41 ± 0.016	0.57 ± 0.022	0.72 ± 0.029	0.91 ± 0.036	1.03 ± 0.041	1.37 ± 0.048
t	p	$t_{1;3} = 5.882$ $p \prec 0.001$	$t_{3;5} = 4.121$ $p \prec 0.001$	$t_{5;7} = 4.111$ $p \prec 0.001$	$t_{7;10} =$ 2.199 $p \prec 0.001$	$t_{10;14} =$ 5.386 $p \prec 0.001$
Z_4	0.83 ± 0.031	1.11 ± 0.033	1.28 ± 0.038	1.45 ± 0.041	1.62 ± 0.047	1.89 ± 0.051
t	p	$t_{1;3} = 6.184$ $p \prec 0.001$	$t_{3;5} = 3.378$ $p \prec 0.001$	$t_{5;7} = 3.041$ $p \prec 0.001$	$t_{7;10} =$ 2.725 $p \prec 0.001$	$t_{10;14} =$ 3.893 $p \prec 0.001$

Table 3.10 Intervals and accuracy of AOD determination by spectral selective mapping of the fluorescence intensity of molecular complexes of CSF polycrystalline films of a human corpse

Parameters	AOD determination interval, T	Statistical parameters	AOD determination accuracy, ΔT
The distribution of the fluorescence intensity of the microscopic image in the "blue" region of the spectrum of optical radiation	24 h	$Z_1 = 0.38 - 0.16$ $Z_3 = 0.52 - 1.45$ $Z_4 = 0.81 - 1.79$	25 min
The distribution of the fluorescence intensity of the microscopic image in the "green-yellow" region of the spectrum of optical radiation	28 h	$Z_1 = 0.48 - 0.11$ $Z_3 = 0.56 - 1.58$ $Z_4 = 0.33 - 1.19$	35 min
The distribution of the fluorescence intensity of the microscopic image in the "red" region of the spectrum of optical radiation	14 h	$Z_1 = 0.54 - 0.23$ $Z_3 = 0.41 - 1.37$ $Z_4 = 0.83 - 1.39$	15 min

An analysis of the parameters (interval and accuracy) of the AOD determination by spectrally selective two-dimensional mapping of the intrinsic fluorescence intensity distributions of various biochemical molecular complexes of CSF polycrystalline films of a human corpse revealed the maximum accuracy of the AOD determination among all of the known methods of laser polarimetry.

3.3 Polarimetric Mapping of the Azimuth of Polarization of Spectral-Selective Microscopic Images of Autofluorescence of CSF Polycrystalline Films

This polarization technique carries other information about the molecular component of the polycrystalline structure of CSF films, which significantly expands the functionality of determining the interval and the accuracy of AOD establishment. The polarization azimuth maps were measured at the location of the Stokes-polarimeter (Fig. 3.9).

Examples of the results of temporary polarization mapping of two-dimensional azimuth distributions of microscopic images of laser-induced fluorescence of protein molecules and nicotinamide nucleotide NADR of CSF polycrystalline films from a human corpse are shown in Figs. 3.10 and 3.11.

Analysis of the data revealed the following:

- Both microscopic images of laser-induced fluorescence of PFCSF are coordinate-inhomogeneous (Figs. 3.10 and 3.11, left parts) regardless of the changes in the observation time after death.

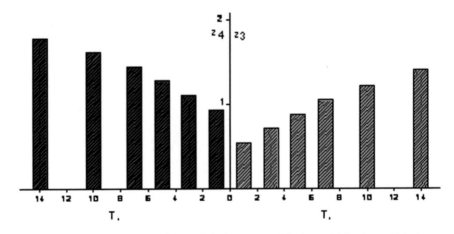

Fig. 3.9 Temporal dependences of the statistical moments of the 1st and 3rd orders, which characterize the distribution of the fluorescence intensities of polycrystalline CSF films in the "red" region of the spectrum

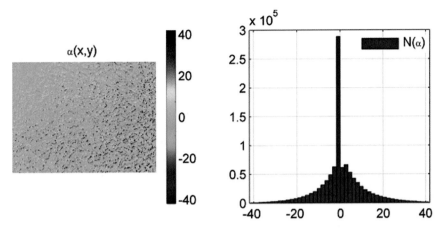

Fig. 3.10 The coordinate structure (left side) and the histogram of the distribution (right side) of the random values of the azimuth of polarization of the microscopic image of the intrinsic fluorescence of PFCSF in the short-wavelength region of the optical spectrum. AOD 1 h

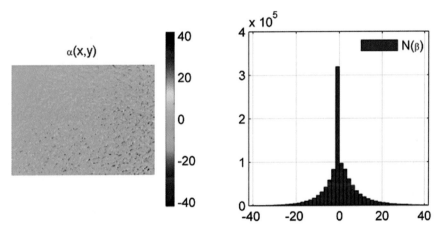

Fig. 3.11 The coordinate structure (left side) and the histogram of the distribution (right side) of the random values of the azimuth of polarization of the microscopic image of intrinsic fluorescence of PFCSF in the short-wavelength region of the optical spectrum. AOD 3 h

- The topological structure of the coordinate distributions of the polarization azimuth is predominantly small-scale and consists of spatially distributed domains, within which various values of this parameter are formed, which represents a certain range of changes in the values of the polarization azimuth in the histograms of its magnitude distribution.
- With increasing observation time due to degenerative changes in the biochemical structure of CSF films of a human corpse, the fluorescence is quenched and the

molecular dipoles are disoriented, which emit in the "blue" region of the optical spectrum.

- A comparative analysis of the histograms of the distribution of random polarization azimuths of laser-induced fluorescence shows that the range of scatter decreases (Figs. 3.10 and 3.11, right-hand sides)—1.4 times in 3 h after death.
- The statistical moments of the 3rd and 4th orders are the most sensitive to degenerative changes, which characterize the skewness and kurtosis (peak sharpness) of the distribution of random values of the azimuth of polarization of the fluorescent radiation of a set of protein molecules and nicotinamide nucleotide NADR.

The results of the calculations of the statistical moments of the 3rd–4th orders, which characterize the coordinate distributions of the polarization azimuth values of the microscopic images of laser-induced fluorescence of CSF polycrystalline films within a statistically significant sample within 20 h after death, are presented in Table 3.5 and Fig. 3.12.

An analysis of the temporal dynamics of the changes in the statistical structure of the polarization map of the azimuths of the microscopic images of the intrinsic fluorescence of CSF polycrystalline films of a human corpse in the "blue" region of the spectrum showed that the range of variation in the values of the 3rd order statistical moment is 2.04 times; for the 4th order statistical moment, the range is 1.88 times (Table 3.11).

On this basis, according to the AOD determination algorithm, the following data were established: AOD detection interval is $T = 20$ h; the accuracy of determining the AOD is $\Delta T = 20$ min.

As can be seen, the interval and accuracy of determining the AOD is similar to the results from the data obtained by polarization-correlation mapping of the time

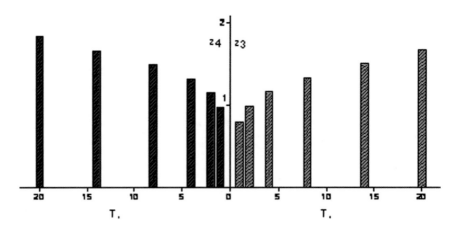

Fig. 3.12 Temporal dependences of the magnitude of the statistical moments of the 3rd–4th orders, which characterize the values of the distribution of the azimuth of polarization of the microscopic image of intrinsic fluorescence of CSF polycrystalline films of a corpse, in the "blue" spectral region

Table 3.11 Temporal dependences of the statistical moments of the 3rd–4th orders, which characterize the distribution of the polarization azimuth values of the microscopic image of laser-induced fluorescence of protein molecules and NADR in CSF polycrystalline films of a corpse

T, hours	1	2	4	8	14	20
Z_3	0.81 ± 0.044	0.99 ± 0.036	1.16 ± 0.032	1.33 ± 0.040	1.5 ± 0.044	1.67 ± 0.049
t	p	$t_{1;2} = 3.929$ $p \prec 0.001$	$t_{2;4} = 3.529$ $p \prec 0.001$	$t_{4;8} = 3.319$ $p \prec 0.001$	$t_{8;14} = 2.859$ $p \prec 0.001$	$t_{14;20} = 2.581$ $p \prec 0.001$
Z_4	0.98 ± 0.025	1.15 ± 0.026	1.32 ± 0.030	1.49 ± 0.035	1.66 ± 0.042	1.83 ± 0.049
t	p	$t_{1;2} = 4.713$ $p \prec 0.001$	$t_{2;4} = 4.282$ $p \prec 0.001$	$t_{4;8} = 3.688$ $p \prec 0.001$	$t_{8;14} = 3.109$ $p \prec 0.001$	$t_{14;20} = 2.634$ $p \prec 0.001$

dynamics of changes in the statistical structure of the microscopic images of samples of CSF films.

When mapping the distribution of the polarization azimuth values of laser-induced fluorescence of CSF polycrystalline films in the "mid-wave" spectral region, a band-pass filter with a transmission maximum $\lambda_2 = 0.55$ mkm was used, which allows one to study the degenerative changes in flavins and folic acids.

Polarization azimuth maps obtained at the location of the Stocks-polarimeter with spectrally selective filtering of microscopic images of CSF polycrystalline films with intrinsic fluorescence change their structure quite quickly within 3 h after death (Figs. 3.13 and 3.14, left parts). Objectively, this process of postmortem degradation of the orientational structure of molecular polycrystalline compounds is illustrated

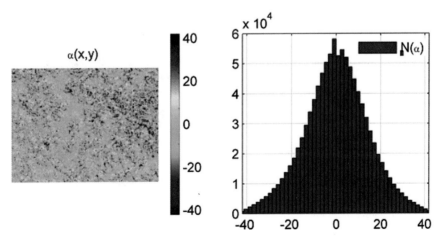

Fig. 3.13 The coordinate structure (left side) and the histogram of the distribution (right side) of the random values of the azimuth of polarization of the microscopic image of intrinsic fluorescence of PFCSF in the mid-wavelength region of the optical spectrum. AOD 1 h

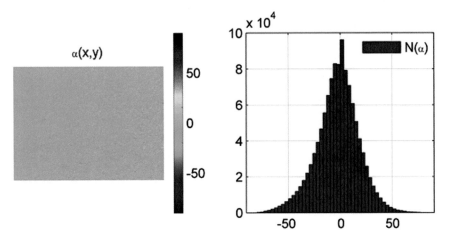

Fig. 3.14 The coordinate structure (left side) and the histogram of the distribution (right side) of the random values of the azimuth of polarization of the microscopic image of intrinsic fluorescence of PFCSF in the mid-wavelength region of the optical spectrum. AOD 3 h

by changes in the dispersion and kurtosis of the scatter of random values of the polarization azimuth in the histograms of the distribution of this parameter (Figs. 3.13 and 3.14, right-hand sides). Statistical data processing within a representative sample shows the values of the statistical moments of the 3rd and 4th orders, which characterize the distribution of the polarization azimuth of the microscopic images of the induced fluorescence of CSF polycrystalline films in the mid-wave region of the optical radiation spectrum for 30 h after death, as presented in Table 3.12 and Fig. 3.15.

Temporal monitoring of the postmortem changes in the statistical structure of the distribution of polarization azimuth values of the fluorescence of the biochemical

Table 3.12 Temporal dependences of the statistical moments of the 3rd and 4th orders, which characterize the distribution values of the azimuth of polarization of the microscopic images of laser-induced fluorescence molecules of flavin and folic acids from CSF polycrystalline films of a corpse

T, hours	1	3	6	14	22	30
Z_3	0.68 ± 0.022	0.85 ± 0.024	1.02 ± 0.029	1.19 ± 0.033	1.36 ± 0.035	1.53 ± 0.041
t	p	$t_{1;3} = 5.222$ $p \prec 0.001$	$t_{3;6} = 4.516$ $p \prec 0.001$	$t_{6;14} = 3.87$ $p \prec 0.001$	$t_{14;22} =$ 3.534 $p \prec 0.001$	$t_{22;30} =$ 3.153 $p \prec 0.001$
Z_4	1.03 ± 0.032	1.21 ± 0.039	1.38 ± 0.042	1.55 ± 0.048	1.72 ± 0.052	1.89 ± 0.056
t	p	$t_{1;3} = 3.568$ $p \prec 0.001$	$t_{3;6} = 2.966$ $p \prec 0.001$	$t_{6;14} =$ 2.665 $p \prec 0.001$	$t_{14;22} =$ 2.402 $p \prec 0.001$	$t_{22;30} =$ 2.225 $p \prec 0.05$

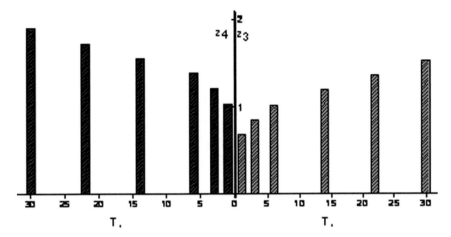

Fig. 3.15 Temporal dependences of the magnitude of the statistical moments of the 3rd–4th orders, which characterize the values distribution of the azimuth of polarization of the microscopic image of intrinsic fluorescence of CSF polycrystalline films of a corpse in the "green-yellow" spectral region

molecules of flavins and folic acids of CSF polycrystalline films revealed ranges of changes in the values of the third-order statistical moment (2.25 times) and the fourth-order statistical moment (1.87) times).

On this basis, the established determination interval of AOD is $T = 30$ h. Moreover, the AOD determination accuracy remains high for $\Delta T = 30$ min in comparison with the data of two-dimensional polarization mapping of fluorescence in the blue region of the spectral range of optical radiation.

When mapping the distributions of the azimuth of polarization of the laser-induced fluorescence of the CSF polycrystalline films in the "longwave" region of the spectrum, we used a bandpass filter with a transmission maximum $\lambda_3 = 0.63$ mkm that allows one to study degenerative changes in porphyrins.

Two-dimensional polarization azimuth maps (left-hand sides) and histograms of the distribution of values of this parameter (right-hand sides) obtained by applying two-dimensional Stocks-polarimetric mapping of intrinsic fluorescence in the "red" region of the spectrum are shown in Figs. 3.16 and 3.17. The data obtained provide additional information on postmortem biochemical changes in CSF at the level of blood content and its components.

A comparative analysis of the data of two-dimensional polarimetric mapping of the intrinsic fluorescence of porphyrins in the red region of the spectrum with similar data obtained in the short and medium wavelength ranges revealed more rapid changes in the two-dimensional and statistical structure maps of the azimuth of polarization at different times of observation after death (Figs. 3.16 and 3.17, left parts).

In Table 3.13 and in a series of diagrams, Fig. 3.18 presents the results of calculations of the statistical moments of the 3rd and 4th orders, which characterize the

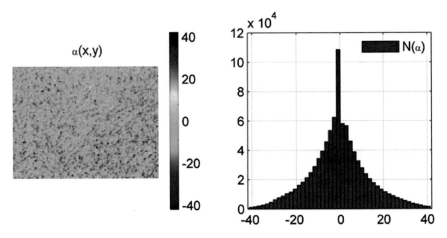

Fig. 3.16 The coordinate structure (left side) and the histogram of the distribution (right side) of the random values of the azimuth of polarization of the microscopic image of intrinsic fluorescence of PFCSF in the long-wavelength region of the optical spectrum. AOD 1 h

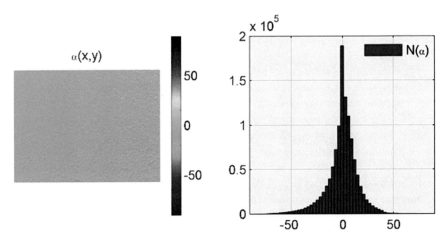

Fig. 3.17 The coordinate structure (left side) and the histogram of the distribution (right side) of the random values of the azimuth of polarization of the microscopic image of intrinsic fluorescence of PFCSF in the long-wavelength region of the optical spectrum. AOD 3 h

skewness and kurtosis of the coordinate distribution values of the azimuth of polarization at the points of microscopic images of the intrinsic fluorescence of polycrystalline networks of porphyrin in CSF films within representative samples of deceased from cardiovascular pathology over 18 h after death.

It has been established that the most distinct and dynamic temporary postmortem changes in the coordinate structure of the polarization maps of the azimuth of microscopic images of laser-induced fluorescence of porphyrins in CSF polycrystalline

Table 3.13 Time dependences of the magnitudes of the statistical moments of the 3rd and 4th orders, which characterize the distribution of the azimuth of the fluorescence of porphyrins in CSF

T, hours	1	4	6	8	12	18
Z_3	0.58 ± 0.017	0.75 ± 0.024	0.92 ± 0.029	1.19 ± 0.035	1.36 ± 0.041	1.57 ± 0.048
t	p	$t_{1;4} = 5.78$ $p \prec 0.001$	$t_{4;6} = 4.516$ $p \prec 0.001$	$t_{6;8} = 5.94$ $p \prec 0.001$	$t_{8;12} =$ 3.153 $p \prec 0.001$	$t_{12;18} =$ 2.693 $p \prec 0.001$
Z_4	0.82 ± 0.037	0.99 ± 0.026	1.16 ± 0.030	1.33 ± 0.037	1.5 ± 0.041	1.67 ± 0.049
t	p	$t_{1;4} = 3.759$ $p \prec 0.001$	$t_{4;6} = 4.282$ $p \prec 0.001$	$t_{6;8} = 3.569$ $p \prec 0.001$	$t_{8;12} =$ 3.078 $p \prec 0.001$	$t_{12;18} =$ 2.661 $p \prec 0.001$

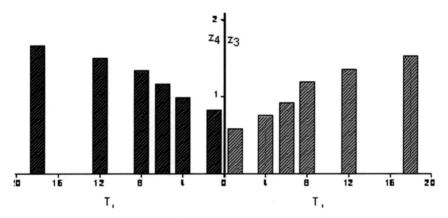

Fig. 3.18 Temporal dependences of the statistical moments of the 3rd and 4th orders, which characterize the distribution of the azimuth of polarization of the microscopic image of the intrinsic fluorescence of porphyrins of CSF films

films affect the values of the statistical moment of the third order (2.64 times); and the range of variation in the values of the statistical moment of the 4th order is 2.04 times.

A comparative analysis with polarization mapping data in other spectral ranges revealed a decrease in the interval of the AOD determination due to a faster increase in the values of the statistical moments of the 3rd and 4th orders, $T = 18$ h. At the same time, a maximum accuracy of AOD determination of $\Delta T = 15$ min is achieved.

The following are the results of determining the intervals and the accuracy of determination of the AOD by spectral-selective mapping of the distribution of azimuths of polarization of microscopic images of fluorescence of fluorophores of CSF polycrystalline films, as in Table 3.14.

Table 3.14 Intervals and accuracy determination of AOD by polarimetric mapping of microscopic images of fluorescence in CSF films

Parameters	AOD determination interval, T	Statistical parameters	AOD determination accuracy, ΔT
Distribution of azimuths of polarization of fluorescence in the "blue" region of the spectrum	20 h	$Z_3 = 0.81 - 1.67$ $Z_4 = 0.98 - 1.83$	30 min
Distribution of azimuths of polarization of fluorescence in the "green-yellow" region of the spectrum	30 h	$Z_3 = 0.68 - 1.53$ $Z_4 = 1.03 - 1.89$	20 min
Distribution of azimuths of polarization of fluorescence in the "red" region of the spectrum	18 h	$Z_3 = 0.58 - 1.57$ $Z_4 = 0.82 - 1.67$	15 min

An analysis of the parameters (interval and accuracy) of the AOD determination by Stocks-polarimetric mapping of the distribution values of the azimuth of polarization of the microscopic images of intrinsic fluorescence of CSF polycrystalline films revealed the maximum level of accuracy determination of AOD among all of the techniques of laser Stokes-polarimetry.

To identify the effect of alcohol on the results of the AOD determination, the relationships between the distributions of the intensity and azimuths of polarization of the intrinsic fluorescence in CSF polycrystalline films of the deceased due to cardiovascular pathology and severe alcohol poisoning and the magnitudes and ranges of the statistical moments of the 1st–4th orders, which characterize such autofluorescence images, were additionally investigated.

A good level of balanced accuracy of differentiation of the cause of death in the "red" region of the spectrum was achieved: $Ac(M_{1;2;3;4}) \leq 90\%$ (Table 6.15). At the same time, the value of the balanced accuracy of differentiation in the case of weak alcohol poisoning is significantly less (10–15%) and does not reach a satisfactory level (Ac $\prec$ 75%), with the exception of the "red" region of the spectrum (Ac $\sim$ 80%).

Conclusions

This monograph theoretically substantiates and experimentally establishes a set of relevant forensic diagnostics of polarization-correlation and spectrally selective relationships between the time and causes of the onset of death as a result of cardiovascular disease and severe alcohol intoxication with possible fatal outcomes. This approach uses the statistical moments of the 1st–4th order, which characterize the temporal changes in the coordinate distributions of the complex degree of mutual polarization of different-scale components of microscopic images, of the intensity and azimuth of the polarization of fluorescence in polycrystalline films of cerebrospinal fluid in the postmortem period.

1. According to the characteristics of the coordinate distributions of the large-scale component of the complex degree of mutual polarization, we first established here the following: the interval for determining the time of the onset of death is $T = 40$ h, and the accuracy of determining the time of the onset of death is $\Delta T = 1$ h, by evaluating the temporal dynamics of the changes in the magnitudes of the statistical moments of the 3rd (from 0.56 to 1.41) and 4th (from 0.59 to 1.38) orders.

2. For the small-scale component of the magnitude of the complex degree of mutual polarization of microscopic images of polycrystalline films of cerebrospinal fluid, the interval for determining the time of the onset of death is $T = 10$ h with an accuracy of $\Delta T = 25$ min, which were determined for the first time by determining the range of the monotonic changes in the skewness (from 1.19 to 2.04) of cerebrospinal fluid and the kurtosis (from 0.91 to 1.86).

3. Forensic medical parameters were established for autofluorescence determination of the time of the onset of death with an accuracy of 15 min at a time interval of 6 h for the statistical moments of the 1st and 3rd orders, which characterize the average and skewness of the distribution of the intrinsic fluorescence intensity of the films of the cerebrospinal fluid in the "red" region of the spectrum and vary in the following limits: average from 0.54 to 0.15; skewness from 0.93 to 1.79.

© The Author(s), under exclusive licence to Springer Nature Singapore Pte Ltd. 2021   65
M. S. Harazdyuk et al., *Correlation and Autofluorescence Microscopy in Forensics Medicine: Time of Death Detection Using Polycrystalline Cerebrospinal Fluid Films*, SpringerBriefs in Physics, https://doi.org/10.1007/978-981-16-0197-2

4. It was established here that the temporal dynamics of the changes in the magnitudes of the statistical moments of the 3rd and 4th orders, which characterize the coordinate distribution of the azimuths of polarization of laser-induced fluorescence in "green-yellow" (skewness from 0.68 to 1.53; kurtosis from 1.23 up to 1.89) and "red" (skewness from 0.58 to 1.57; kurtosis from 0.82 to 1.67) sections of the spectrum, ensure the accuracy of determining the time of the onset of death to be 15 minutes at a time interval of 6 hours after the onset of death.

5. Reliable signs have been identified for the forensic differential diagnosis of cardiovascular pathology and severe alcohol intoxication with a possible fatal outcome by the method of laser-induced fluorescence in the "red" region of the spectrum, with excellent balanced accuracy Ac (M1, 2, 3, 4) > 90%.

Printed in the United States
by Baker & Taylor Publisher Services